Happy Birthday

Love & Best Wishes Always.

Rita. xxxx 1978.

FAST
ATTACK CRAFT

Also by Martin H. Brice:

The Tribals *(Ian Allan Ltd 1971)*
The Royal Navy and the Sino-Japanese Incident 1937-41 *(Ian Allan Ltd 1973)*

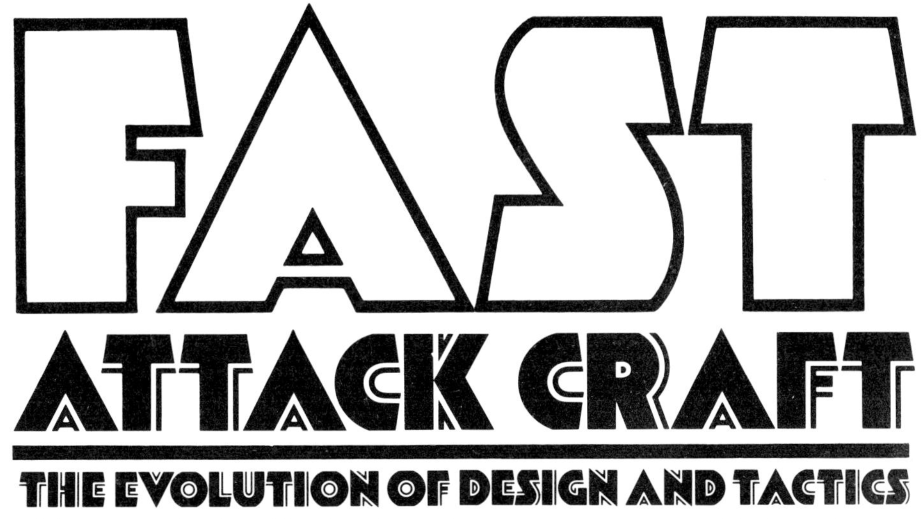

KEIREN PHELAN & MARTIN H. BRICE

MACDONALD AND JANE'S LONDON

First published in Great Britain 1977 by
Macdonald & Jane's Ltd, Paulton House,
8 Shepherdess Walk, London N1 7LW

ISBN 356 04912 4
Printed in Great Britain by Netherwood Dalton & Co. Ltd, Huddersfield

CONTENTS

ACKNOWLEDGMENTS

Among the many friends and correspondents who have provided information and encouragement we should like in particular to thank the following: A. Augustin-Normand, P. Boyd, Len H. Cox, Lt Cdr P. Dalzel-Job RN (Rtd), Kevin Desmond, the late R. I. N. Douglas, Cdr Peter Du Cane CBE RN (Rtd), Cdr Sir John Eardley-Wilmot MVO DSC RN (Rtd), the late Lt Cdr R. I. T. Falkner, Weston Farmer, Jack Garside, Edwyn Gray, Eric Green, D. E. J. Hunt, John Liley, Anthony Needell, John Nicholson, Anthony Preston, John Pritchard, Capt E. N. Pumphrey DSO DSC RN (Rtd), Frank H. Snoxell, William C. Sheaff, Michael B. Stevens, Vic Stride, Sheila Watson, Bill Wilson.

We are also indebted to the staff and resources of Alton Public Library, Bibliothek Für Zeitgeschichte, Stuttgart, Imperial War Museum, Liddell Hart Centre for Military Archives (University of London King's College), National Maritime Museum, Ministry of Defence Naval Library, Newspaper Library, Colindale, Portsmouth City Library, Public Record Office, Royal Institution of Naval Architects, Royal United Services Institute, Science Museum Library.

'Blinded by the search-light, their attention occupied by the supposed Russian, none of the enemy saw us steal past until it was too late for them to interfere with our attack.

We had passed them; and ahead of us lay two long lines of battleships, some looming up black and silent in the darkness, others lit up by the flashes of the random guns they had already begun to fire. Mostly they used no search-lights, fearing thereby to bewilder their gunners; nor as yet were they certain in which direction to look. Before long, however, a chance beam from the electric light fell upon our leading torpedo boat, now going for between the lines at her utmost speed, the dummies trailing well out astern of her. I suppose the Russian flag puzzled the French, for there was a lull in the firing, some signalling, shouting, and momentary indecision; but as No. 87 reached the farther end of the lines, a vigorous cannonade began again, and at the same instant we heard the detonation of a couple of torpedoes. The psychological moment had come!

Blake blew his signal whistle; we tore down the Russian flag, hoisting the white ensign in its place; and off we went between the shore and the enemy till we had passed the last of them; then, circling round, we charged back to complete the work of destruction we had begun. Scarcely a shot came near us, as our torpedoes went home one after the other with a series of the most awful explosions I have ever witnessed. By the time we had been up the lines and down again – a space of but a few minutes – of all that great armada but two ships were left firing; all the rest had sunk or run aground!'

Blake of the 'Rattlesnake' *by Fred T. Jane 1895*

INTRODUCTION

In Ben Jonson's comic play *The Staple of News* the object of his satire is the unlikely fiction created by the newsmongers of early seventeenth century London. As this extract shows, Jonson's fantasy was remarkably prophetic.

Thomas: They write here one Cornelius – Son,
Hath made the Hollanders an invisible Eele,
To swimme the Haven at Dunkirke, and sinke all
The shipping there.
Peniboy: But how is't done?
Cymbal: I'll shew you sir.
It is an Automa, runnes under water,
With a snug nose, and has a nimble taile
Made like an auger, with which taile she wrigles
Betwixt the coasts of a Ship, and sinks it streight.
Peniboy: Whence ha' you this newes?
Fitton: From a right hand I assure you,
The Eele-boats here, that lye before Queen-Hyth,
Came out of Holland.
Peniboy: A most brave device,
To murder their flat bottoms.

Act III scene ii, 1631

It was the evolution of the torpedo, a self-propelled explosive underwater weapon, which led directly to the introduction of fast offensive boats in the 1860s and 1870s. The torpedo came to notoriety during the American Civil War when boats and submersibles carrying explosive charges at the end of long wooden spars made clandestine attacks on moored shipping. These vessels were far less successful than widely supposed; the real slaughter came from mines. Russia had used mines for harbour defence during the Crimean War, to little effect, but the havoc caused by controlled and contact mines during the American Civil War led to panic in European navies, causing a rash of defences against the torpedo, most of them useless. (For many years spar torpedoes, mines, and other brave devices were all known generically as torpedoes, until the introduction of the self-propelled, fish, or automotive torpedo itself led to accurate terminology.) When rigged with a spar torpedo, all manner of naval cutters and pinnaces posed an impish threat to shipping, and one most effectively countered by other spar torpedo boats. Consequently they became lightly armed and the fastest boat was the one most likely to return unscathed to its harbour or parent ship. Torpedo boats became potent after 1879 when they began to carry Whitehead's torpedo, and were built with increased size, speed, and seagoing ability in an attempt to enable them to carry the war to the enemy and over many years they occasionally fulfilled their menace. Then, as now, a determined, bold, and experienced crew with a well-found boat could sink a capital ship. Nineteenth century torpedo boats could close the enemy, sometimes at will and frequently with impunity, and there was a surge of torpedo boat construction in the 1880s. Most of the world's torpedo boats were built in Great Britain, the country which introduced the concept but had less strategic use for it than any other nation except America, which soundly refused to indulge in the raging naval fashion.

The rapid growth in size of torpedo boats was the inaugural example of a recurring disposition in fast fighting boats. To improve all aspects of their performance fast boats were increased in size until they reached the stage where they in turn became targets for impudent attacks by daring men in fast attack craft. In the late nineteenth century torpedo boats were classified as second class (about 18·2m [60ft] in length and carried aboard cruisers and battleships or used for harbour defence); first-class of 38·1m (125ft) or so, and sea-going, of over 42·7m (140ft). The expansion of rôle, though not necessarily capability, led directly to the introduction of the destroyer, a warship which was frequently the target of fast attack craft in both world wars.

The exception to this congenital bloating was the 12·2m (40ft) coastal motor boat, introduced in 1916 for the specific purpose of firing its torpedo inside the German minefields laid off North Sea ports. A 16·8m (55ft) version was also produced but it was still a sufficiently small, intimate craft for its crew not to lose sight of their operational function, difficult, uncomfortable and dangerous though it might be. Probably the most triumphant instance of small boat warfare was the work carried out with a handful of coastal motor boats in the Baltic in 1919, when the qualities of the boats were utilised for specific purposes. The Italian navy has been alone in both world wars in exploiting consistently and successfully the zeal, resolution, and self-reliance forged by the camaraderie aboard small craft in action.

One of the traditional arguments in favour of fast boats, trumpetted at reluctant admiralties, was that a dozen, a score even, could be built for the cost of one warship. This naturally carried a great deal of weight among smaller navies, and it might be imagined that with the excruciating cost of modern naval systems this aspect is still relevant. However two factors have altered the condition of fast attack craft being principally the weapon of weaker naval powers. One is the diminishing ability of surface warships to detect and destroy submarines, which considerably reduces their validity in modern navies. The other is the enormous growth in the last fifteen years of the number of fast attack craft equipped with missiles. The fast attack missile craft is on an equal footing, and frequently a superior one, to an equivalently armed warship, far more so than were torpedo boats even in their most fearsome days. The missile build-up started in the late 1950s, when Russian Komar-class boats were fitted with two Styx missiles, long before the formidable striking power became apparent in the Arab-Israeli and Indo-Pakistan Wars. Electronic warfare has made redundant many aspects of small boat combat (the Israeli destroyer *Eilat* was sunk in 1967 by Styx launched from a Komar boat moored 10-13 miles away in Port Said harbour) and there is little which can be done to counter 'a sea-skimming missile, either homing on infra-red emissions or guided by a sophisticated, near-unjammable radar seeker, fired from 25 to 50 miles away by a small elusive platform which may not even appear on the target ship's radar screen.'[1] In narrow or disputed waters, though, these awesome weapons promote both stealth and high-speed seamanship, and their value to numerous navies is seen in the large numbers employed. Russian missile attack craft, notably Osa I and Osa II classes, are in widespread use with Warsaw Pact countries and other Soviet client nations. Russia herself operates 120 missile attack craft, and some 230 fast torpedo and gun boats, of which 40 are hydrofoils. China also has 120 missile boats, in addition to about 700 other fast attack craft; most of this swarm, the core of the Chinese navy, are Russian built boats, though since the Sino-Soviet rift China has been building fast gunboats of her own design to police her tortuous coastline. Even in times of peace, small combatants are no longer just the backbone of lesser powers but increasingly represent the ungloved fist of all navies concerned with preserving the security of their waters. Greece and Turkey each have missile boats, as do Egypt (18), Israel (18), and Iraq (10). All the Baltic nations operate them; Norway has 26. At present some 490 fast missile attack craft are in service around the world.

Most of the West's craft are built in France, Germany, and Britain by a handful of countries. With *La Combattante,* built by Chantier Mécaniques de Normandie at Cherbourg, France has regained an eminence not seen since the Normands of Le Havre rivalled the builders on the Thames in the nineteenth century. In Germany Lürssen continues to produce excellent fast patrol boats and in Britain Brooke Marine of Lowestoft are competing strongly in the large market for overseas orders long dominated by Vosper Thornycroft. European shipyards lead the world in the provision of seaworthy cost-effective attack boats for the protection of narrow, and not so narrow, seas. Germany operates 40 fast attack missile and torpedo craft, but France, Britain and Italy have less than a dozen such boats between them. It would be interesting to consider the benefits a single European class of fast attack craft might bring to these navies, who for reasons of cost are unable to support vessels for which there is no immediate operational necessity. Already the constituents of fast attack craft are thoroughly international. A boat built in France may have German diesels or British gas turbines, Dutch fire control, Swedish or Swiss guns, Italian or French missiles. Lürssen and CMN started a programme of cooperation in 1970 to produce the Type 148, to replace Germany's withering flotillas of fast patrol boats stationed in the Baltic.

The Royal Navy's interest in light coastal forces has waxed and waned for a hundred years, enthusiasm naturally falling off after the building sprees caused by war. The Coastal Motor Boat branch was paid off in 1926, and the dissolution of CMB experience was dearly paid for in the opening years of the Second World War.After the war the axe did not fall again until 1957, when interest in coastal forces was officially concluded. However, while the relatively straightforward boats of the Second World War could be designed and built in crash programmes, the complexities of today's hyperbolic weapons systems mean that there can be no picking up of threads, and it is disturbing that the Royal Navy's three high-speed training craft completed in 1970 are no longer in full commission. Britain could readily design, power, construct, and arm suitable fast attack craft – it is already doing so for other navies – but it could not provide the tactically trained crews to operate them with any hope of success. The essence of successful fast boat actions has been a positive and predetermined assault, a sound boat, and an experienced and well-led crew. Light forces have in the past been manned in wartime largely by naval reserves and volunteers, usually those in sympathy with the way of small ships at sea;

currently there is pitifully little expectation of there being personnel to utilise the exacting performance of a fast attack craft built by this country or a European consortium.

The dimensions given in this book, metric followed by imperial, refer throughout to, first, overall length (unless stated to be waterline or between perpendiculars), second, beam, and third, draught. The tonnage given is displacement unless otherwise stated.

Even where maximum and maximum continuous speeds are given, it should be remembered that the figures are not definitive because so many factors influence, usually adversely, a boat's supposed speed. Performance logged on acceptance trials may never again be achieved in a boat's lifetime; conversely a crew could often exceed a vaunted maximum, by a margin dangerous to the machinery. The speeds given are generally theoretical maximums, but equipment, extra stores and fuel, or a duff engine could reduce the ordained top speed by 5 knots or more. Besides, fuel consumption at maximum revolutions is so great that a boat's highest speed was wisely best attained in the wardroom. Armament also frequently differed from that originally fitted, sometimes as a result of sanctioned improvements, more commonly as a result of trying to fit bigger or better guns.

Individuals have been singled out for their accomplishments in the design of fast boats but it should not be forgotten that, particularly during wartime, these achievements were also the achievements of many others, working with and for them. The accomplishment of this book is the combined work of two authors, in which Keiren Phelan wrote the first half and Martin H. Brice the second.

1 TORPEDO

Scouring naval chronicles for the origins of the torpedo indicates that floating bombs, of variable ingenuity and doubtful reliability, have been in use for many centuries. A random, but influential, torpedo attack was Zambelli's floating mine of 1585, an innovation used by the Dutch against a Spanish-held bridge over the Scheldt during the siege of Antwerp. Boats packed with gunpowder were carried by the tide to the bridge, which was destroyed when the powder was fired by clockwork-operated flintlocks. This triumph may have been responsible for the Dutch enthusiasm for fireships in the seventeenth century. During the War of American Independence (1775-83) a timid attack on English ships moored in the Delaware was made in December 1777 by means of barrels of gunpowder drifting downstream towards the ships at Philadelphia. The barrels were submerged but the buoys to which they were attached were hugely visible. The English ships had been warped into the docks to avoid ice, but seeing the barrels the men ran to the wharves and opened fire on them. For this the action was dubbed the Battle of Kegs – the crew of one ship's boat was blown up though when it attempted to recover one of the barrels.

The idea of drifting mines, designed to explode when jarred by contact, came from the American David Bushnell (1742-1824). He experimented with numerous explosive devices and succeeded in showing that an underwater explosion would not be absorbed by water but was contained by it. During his residence at Yale College Bushnell was busy trying to devise a means of underwater attack and, with the incentive of the American War of Independence, designed and built a one-man submarine. With this egg-shaped weapon (driven by screw-propellers years before the propeller was 'invented') he planned to fix a charge of 68·1kg (150lb) of gunpowder to the hull of a ship. The *Turtle* was convincingly successful but failed as a weapon; when trying to screw the charge to the bottom of the frigate HMS *Eagle*, it was found that *Turtle* was not sufficiently buoyant to force the auger through the hull but the charge was exploded during the submarine's retreat. Bushnell had a greater success in August 1777 when a mine was hauled aboard a prize schooner taken by HMS *Cerberus* at New London, Connecticut. Three curious seamen goaded it into action and it exploded, killing the sailors and sinking the schooner. It was an ingenious device judging from its description in a letter sent by Commodore Symons *(Cerberus)* to Rear-Admiral Sir Peter Parker (15 August 1777), and fastened with a long line, it anticipated the towed torpedo introduced by the Harveys in 1869. For all his creativity Bushnell's martial designs were not successful, and General George Washington wrote to Thomas Jefferson (26 September 1785): 'One accident or another always intervened. I then thought, and still think, that it was an effort of genius, but that too many things were necessary to be combined to expect much from the issue.'

Twenty years later Robert Fulton (1765-1815) had started to experiment with torpedoes. Born in Pennsylvania, he went to England in 1878, and with his energy and enthusiasm he eventually fell in with Lord Stanhope, Robert Owen and Edmund Cartwright. Like these resourceful men, Fulton was brimming with ideas which suffered, however, from want of foresight. His digging machine was quite sound but only to the extent that it was for excavating the small canals he favoured, which could carry boats only 1·5m (5ft) wide. Although regarded as an inventor, his ideas often lacked originality. He patented in 1794 an inclined plane for use on Stanhope's Bude Canal for moving boats from one level to another, but learned from Stanhope that a superior version had been conceived by Edmund Leech in 1777. Moreover he did not arrive in France to start work on his submarines and torpedoes until two years after a pamphlet describing Bushnell's experiments had been published in Paris. It is for his work in applying steam power to boats that Fulton is most respected and before his death he designed at least nineteen steam vessels, from 23·8m to 53·3m (78 to 175ft) in length. These included the *Clermont,* which marked the successful introduction of commercial steam navigation (1807), and the *Demologos* (renamed *Fulton*), which was the world's first steam-powered warship. Conceived during the 1812-14 War as a harbour defence vessel for New York, but not operational until 1815, she was a twin-hulled craft, with a paddle wheel operating between the hulls, and averaged 4·7 knots. Fulton was an enterprising engineer when the project in hand stretched his talents.

In 1797 he took advantage of the brief armistice between England and France to visit Paris, hoping for better luck with his ideas on canal construction and navigation. Whether it was the continuing failure of his canal schemes, the English blockade of Channel ports, meeting Bushnell, or seeing Bushnell's pamphlet, or vigorous republican ideals is hard to decide, but before the end of 1797 Fulton had experimented with torpedoes on the Seine. A design, as he phrased it, 'to impart to carcasses of gunpowder a progressive motion under water to a given point, and there explode them'. On 13 December he wrote to the French Directory outlining his terms for the construction of a 'Mechanical Nautilus' capable of annihilating the English Navy. Fulton was convinced, with some justification, that once his idea had been put into successful practice, its power as a terrible deterrent would be so great that no nation would dare war with another and freedom of the seas would ensue. After showing a languid interest, the Minister of Marine, Pléville-le-Pelley, rejected the proposals on 5 February 1798. With his second supplication, six months later, Fulton had more success. Bruix, a more amenable minister, appointed a commission of eight men and the sense of their report,[1] delivered on 5 September 1798, is a tribute to its subject and their ability to foresee its value.

The report describes and illustrates Fulton's *Nautilus*, a propeller-driven submarine 6·5m (21ft 3in) in length with the ability to attach 45·5kg (100lb) 'submarine bombs' to ships' hulls and detonate them. The commission suggested that with modifications *Nautilus* should be manufactured but the Directory was unmoved by the project and it was again rejected. When Bonaparte muscled his way into power on 9 November 1799, the Directory was abolished and Forfait (a member of the commission) became Minister of Marine, and in the spring of 1800 construction started on *Nautilus*. She was launched at Rouen on 24 July and successfully tested in the Seine and one week later was towed to Havre for sea trials. From there Fulton took the boat on a hazardous sea voyage to Cherbourg, and later (probably overland) to Brest.

While at Brest where *Nautilus* was being refitted, Fulton arranged with the maritime prefect, Caffarelli, the construction of an archetypal torpedo boat: a small vessel, faster than its enemy and loaded with torpedoes. It was an 11·0m (36ft) pinnace, powered by 24 seamen turning four cranks driving a propeller. Though bestowed with official approbation, the boat only achieved 4 knots instead of the 12 hoped for and the plan was dropped. He also built a model of a 300-ton vessel which was to have sides 1·8m (6ft) thick with 14·7cm (6in) decks and be propelled by a manually operated propeller, probably similar to the arrangement in the 11·0m (36ft) pinnace, though larger. Its interesting feature is the early, if not the first, appearance of the spar torpedo, in which a canister of explosive is prodded at the enemy by means of a long pole. Fulton had big bombs in mind on this project because the spars, four projecting over each quarter, were 29·3m (96ft) long, supported by guys from the masthead.

When he was satisfied with the efficiency of *Nautilus,* Fulton

'quit the experiments on the Boat to try those of the Bomb Submarine. It is this bomb which is the Engine of destruction, the plunging boat is only for the purpose of conveying the Bomb to where it may be used to advantage. . . The Prefect Maritime and Admiral Villaret ordered a small sloop of about 40ft [12·2m] long to be anchored in the Road [at Brest] on the 23rd Thermidor [August 11]. With a bomb containing about 20lb [9·1kg] of powder I advanced to within about 200m [200yd]; then taking my direction so as to pass near the sloop, I struck her with the bomb in my passage. The explosion took place and the sloop was torn into atoms, in fact, nothing was left but the buoy and cable; and the concussion was so great that a column of water, smoke and fibres of the Sloop were cast from 80 to 100 feet [25 to 30m] in the air. This simple experiment at once proved the effect of the Bomb Submarine to the satisfaction of all spectators.'[2]

But not to the Consulate, for Forfait resigned in October and the new minister, De Crès, was one of the majority who considered that submarine warfare was sly and horrendous and its retributions would be calamitous, if not divine. Accordingly Fulton abandoned torpedo warfare in France.

After successfully running a steam boat on the Seine, Fulton returned to London in May 1804 to buy a Boulton & Watt engine. During its construction he proposed to the British ;overnment plans for the destruction of French vessels not dissimilar to plans lately made for the destruction of English vessels. Napoleon was amassing his flat-footed invasion fleet at this time and Fulton's submarine was one of the *backfisch* schemes taken up by a panicking government. A commission (including William Congreve and John Rennie) disliked the submarine but felt that the torpedoes might just do, and the misconceived Catamaran Expedition was created.

Fulton constructed a number of oblong coffers, lined with lead and packed with gunpowder, which were to be towed by one-man catamarans (perhaps of the seaside pedalo variety) towards a ship to be fastened by a grappling iron around its anchor line. The black-clad seamen then paddled back to their mother ship (shepherded by Lord Keith in HMS *Monarch*) pulling out a pin to start a clockwork detonator. The attack was made on 2 October 1804 in Boulogne Harbour but only one bomb exploded though it dutifully destroyed a French pinnace. Another attempt was made on 10 December at Calais and the final fling was on 30 September 1805 at Boulogne, both attacks being less successful than the first. The French thought the attacks ridiculous (an obligatory feeling as they had been given first option on the idea) and they were even lampooned and denigrated in England, though they were at least more offensive than the Royal Military Canal or Brigadier-General Twiss's martello towers. To demonstrate that the operation, as well as the principle of torpedoes was sound, Fulton blew up a 100-ton brig with an 81·7kg (180lb) torpedo under her hull, to the huge satisfaction of the large crowd. The experiment was made on 15 October 1805 in Walmer Bay off Deal in Kent and the destruction of the *Dorothea* reinstated Fulton in naval and armament circles, though Earl St Vincent said that Pitt 'was the greatest fool that ever existed to encourage a mode of war which they who commanded the seas did not want, and which if successful would deprive them of it'.

Twelve days later another torpedo attack was made upon two French brigs in Boulogne Harbour. In this attack, which differed in the technique used to destroy the *Dorothea,* a weighted torpedo drifted against the vessel and the pressure of tide carried the torpedo under the bottom of the hull where it exploded. In the third French attack the torpedoes were of a floating variety. Although the torpedoes exploded, they did so on the surface and, as a result, the only injury to the French (five dead, eight wounded) was from small arms fire. Fulton later devised a harpoon for firing torpedoes at ships, though still operated at the end of a line which allowed the torpedo to drift under the hull.

Not long after this, news of Nelson's success at Trafalgar reached the nation, and the necessity for Mr Fulton and his irregular warfare was obviated. He returned to America at the end of 1806 to considerable acclaim and repeated his party piece by destroying a vessel anchored in New York Harbour (on 20 July 1807). Congress was impressed as Fulton intended, for he wrote:

'I consider it a fortunate circumstance that this experiment [*Dorothea*] was made in England, and witnessed by more than a hundred respectable and brave officers of the Royal Navy; for, should Congress adopt torpedoes as a part of our means of defence, Lords Melville, Castlereagh, and Mulgrave, have a good knowledge of their combination and effect.'[3]

(Too great a knowledge really, because Commodore Sir E.W.C.R. Owen, who led the second Boulogne flotilla, was able to write a report to the Admiralty on 6 September 1807[4] describing the torpedoes and a method of coping with them, by keeping the lower studding sail booms out at night and passing lines from the boom ends to a point ahead of the anchor cable, which would prevent torpedoes drifting into the ship.)

In another trial at the end of 1811 Fulton failed to penetrate the molly-coddling defences of Commodore John Rogers' sloop *Argus,* whereupon even America lost interest in torpedoes. Fulton wrote in his defence:

'The nets, booms, kentledge, and grapnels, which he [Rogers] arranged around the *Argus,* made at first sight a formidable appearance against *one torpedo boat and eight bad oarsmen...* I might be compared to what Bartholomew Schwartz, the inventor of gunpowder, would have appeared had he lived at the time of Julius Caesar, and presented himself before the gates of Rome with a four-pounder, and endeavored to convince the Roman legions that with such a machine he could batter down the walls and take the city: a few catapultas, casting arrows, and stones, would have caused him to retreat; a shower of rain would destroy his ill-guarded powder, and the Roman centurions would therefore call his machine a useless invention, while the manufacturers of catapultas, bows, arrows, and shields, would be the most vehement against further experiments!'[5]

The War of 1812-14 brought on a rash of patriotic attempts to sink English vessels with torpedo weapons, the most noteworthy being the half dozen efforts of Mr Mix to blow up HMS *Plantagenet* in Lynn Haven Bay. Two other attempts involved specially built craft of technical interest: one was produced by Fulton in 1814 and by virtue of its shape was known as a turtle boat. Massively constructed, with the bulk of the hull submerged, it was propelled by a dozen men cranking two paddle-wheels, achieving 3·4 knots. The vessel was able to tow five torpedoes and when a resistence on the tow was felt the line was tugged and the torpedo detonated alongside its target.[6] The other effort was by an inventive wizard called Berrian, to whom must be accredited the first spar torpedo boat to be built, though it ran aground before it saw action and, like Fulton's turtle boat, was destroyed by the English. She had an armoured deck, with machinery capable of discharging three large bombs and at 'one end of the boat projected a long pole under water, with a torpedo fastened to it, which, as she approached the enemy in the night, was to be poked under the bottom of a 74, and then let off'.[7]

One of Fulton's inventions was adopted with more vigour but less fuss than the torpedo. This was perhaps because of its passive nature, for even in its crude form Fulton's anchored torpedo was a good defensive weapon; it was the first moored mine, a copper canister, packed with 45·4kg (100lb) of powder and a trip firing device, which was tethered to the sea bed and floated 3m to 5m (10 to 15ft) beneath the surface. Fulton had even considered electricity as a detonator but thought it less practical than his mechanical means.

With the end of the war and Fulton's death, interest in torpedo science and warfare waned until 1829 when Samuel Colt (1814-62) started the first of many successful experiments on electrically operated mines for harbour defence. On 4 July 1829, at Ware Pond, USA, he remotely blew up a raft from land, with 'electric fluid' to detonate the charge. By 1841 he had perfected his system of harbour defence and applied for a government grant and an annuity. He was able to continue his experiments under the authority of the Secretary of War and on 4 June 1842, he successfully demonstrated in New York harbour that a mine could be exploded under water by electricity. On 4 July, he destroyed the US gunboat *Boxer* to considerable acclaim and on 20 August he sank a schooner on the Potomac River some five miles from his position on the land. On 18 October the 300-ton brig *Volta* was blown up at New York and on 13 April of the next year, 1843, the most demonstrative of Colt's experiments took place on the Potomac when a 500-ton brig, manned and sailing at 5 knots, was completely destroyed by one or more electrically operated submarine mines (the crew abandoning ship just before passing into the danger area). The success of these weapons depended on adroit observation and a strategic underwater location of the torpedoes. These provisions were mastered by the Confederacy when it employed torpedoes in the Civil War, seventeen years after Colt's last experiment. Though better known for the invention of an efficient revolver, Colt maintained that he was far more interested in perfecting a system of submarine defences.

At the same time in England Captain S. A. Warner was experimenting with weapons he termed the long range and the invisible shell. The latter was a towed torpedo, containing quite a small charge, and on its only public demonstration it blasted a hole right through the hull and decks of a large barque, *John o' Gaunt,* off Brighton in 1840. The assembly of naval observers was impressed but sceptical and because Warner was disinclined to give any information about his inventions the feeling was that he could have been a trickster. He petitioned for two grants of £200,000 but continually mis-managed his case and by 1844 the project had evaporated.

In the Schleswig-Holstein war (1848-51) mines were successfully used to keep the Danes out of Kiel Harbour but the design of contact mines by the Russians led to their planned and systematic use. This new design was a result of the development of a land mine and can be attributed to Professor Jacobi and Alfred Nobel. It had a chemical fuse which was detonated on being struck by a ship. The Russians also had Colt-type mines, fired electrically from the shore, at Russell, *The Times* correspondent, reported after finding a hull containing batteries and many miles of wire. Russia drew the first blood with mines during the Crimean War (1854-56) though the punch was delivered in the Baltic, during the defence of the island fortress of Kronstadt. HMS *Firefly* and *Merlin* were both badly damaged off Kronstadt (a debt to be repaid in a dashing mission 65 years later) but, once alerted to the danger, Admiral Dundas's fleet was able to disarm many mines.

Finally the American Civil War showed to the world the awful power of mines and their terrible progeny, torpedoes. In its wide use of these weapons the blockaded Confederacy set a style of naval strategy that was copied by countries until the end of the century.

At the outbreak of the war the Federal states had a solid enough navy but one which was poorly equipped to enforce the blockade it imposed on the shallow sounds and rivers of Southern ports. The Confederacy was virtually without a navy and it was that fact which led to such successful improvisation in defending its rivers and ports. The Confederate states' submarine battery service was established to design and develop mines and three main types evolved from the bureau's astonishing efforts. Charleston, Mobile, and Wilmington were all shielded with frame torpedoes: a lattice of five to fifteen mines embedded on the bottom of the river or creek, each mine containing about 12·3kg (27lb) of powder, and virtually impossible to force. After the fall of Charleston in 1865 (it was taken by Sherman's army) the Federal gunboat *Jonquil* was nearly destroyed by a frame mine which exploded during attempts to remove the installations. Moored mines, such as Singer's and Brooke's, were used extensively and towards the end of the war electrically controlled ground mines were used, with devastating effect on one noted occasion. The US gunboat *Commodore Jones* was advancing up the Appomattox River on 6 May 1864, when a boiler containing powder (estimated at between 454kg and 906kg [1000 and 2000lb]) was detonated as she passed over it. The ship simply dissolved into the river, with forty dead and three-quarters of the crew wounded. Of the three men who were seen running from the battery on the shore, one was shot and the other two were captured.

The Federals were aggrieved at the use, or success, of the weapon, describing it as 'asassination in its worst form', 'infernal machinations of the enemy', etc. The South was only defending itself, and when the Federals' hold on North Carolina was threatened in September 1863 they placed a battery of mines in the Roanoke River.

Meanwhile the dramatic slugging match between the ironclads *Monitor* and *Merrimac* had occurred on 9 March 1862. It supposedly led the Irishman John Holland to design his first submarine, thirteen years later. Equally, the appearance of Ericsson's low-slung Federal *Monitor* could have inspired the Charleston boatbuilder T. Stoney to build *David,* in 1863. Many Confederate boats, pickets, launches, and pinnaces were fitted with makeshift spar torpedoes, while ironclads such as *Atlanta* and *Charleston* had heavyweight ram-like torpedoes, though there is no account of the latter's use in action. The spars were from 6·1m to 9·1m (20ft to 30ft) long and most were run out over the stem-head, although the desirability of having two torpedoes led to the practice of carrying a spar on either side of the boat. The 21·3m (70ft) screw tug *Gunnison* was acquired by the CSN in May 1861 and fitted with 68·1kg (150lb) of powder on a spar over the stem, but she never saw action as a torpedo boat.

A small Confederate steamer, *Torch*, was converted into an ironclad, and was also rigged to carry torpedoes. She had a tripod arrangement projecting over her bow bearing three 45·4kg (100lb) torpedoes and on 21 August 1863 she assailed the ironclad *New Ironsides*, the first torpedo boat attack. However she was unwieldy and her engine stopped in the middle of her offensive and she spent the rest of the war as a floating battery in Charleston.

Stoney's 15·2m (50ft) *David* was undoubtedly the first purpose-built steam torpedo vessel. She was constructed of boiler iron, with a beam of 1·8m (6ft) and draught of 1·5m (5ft). She sat very low in the water with just 15·2cm (6in) of freeboard and was cantankerous to handle, but she could puff along at 7 knots. Her main advantage was that she carried her torpedo submerged and positioned for firing at all times. No luckless engineer had to run out the spar and submerge it with weights; everything could be managed from within the hold. On her first mission, commanded by Lieutenant W. T. Glassell CSN, she attacked *New Ironsides*. Her torpedo exploded beneath the starboard quarter but the plumes of water from the explosion doused her boiler fire. Glassell thought that *David* was sinking and under a shower of small-arms fire abandoned her with two others. Her pilot, W. Cannon, remained aboard and was later rejoined by the assistant engineer J. H. Tomb. He managed to get up steam again and *David* was extricated and steered safely back to Charleston Harbour. The ironclad did not sink but was severely damaged. It is not known how many other Davids were employed – perhaps twenty – but they caused fear rather than damage and did not affect the outcome of the Federal blockade.

At least one David was constructed with water-ballast tanks and could make small dives to approach undetected. It was such a boat, with an eight-man crank driven propeller, which sank the Federal steam frigate *Housatonic* off Charleston Harbour on the night of 17 February 1864. The attacker also went down with the frigate, but it was the biggest accomplishment of the war by a torpedo-carrying vessel. This success, as much as anything else, led to the general belief that Davids were submersibles, which was not the case; with this one known exception they were surface craft. On 6 March 1864, *David* tried to sink the blockading Federal ship *Memphis* on the North Edisto River, South Carolina, but twice her torpedo failed to explode and she retreated with a damaged funnel when *Memphis* opened fire. During 1864 four wooden torpedo launches, *Squib, Hornet, Scorpion,* and *Wasp* were built for the Confederates at Richmond. They were 14m (46ft) in length, with a beam of 1·9m (6ft 3in) and draught of 2·1m (6ft 9in). Armed with percussion torpedoes at the end of 5·5m (18ft) spars, they carried out light duties. When *Squib* exploded her charge against the hull of the flagship *Minnesota* (9 April 1864), the torpedo was not sufficiently far beneath the hull to do serious damage and none of the launches caused any real upset to the North. On 19 April 1864 *David* tried to bite the blockading frigate *Wabash* off Charlestown, but could only snap as *Wabash* saw her coming and moved away - and that was the last piece of action *David* saw and the end of the South's torpedo boat warfare.

During this flurry of activity the Federal government realised that torpedo boats were a small price for the rebels to pay for sinking or even damaging an ironclad. The North had fitted spar torpedoes to some tugs and gunboats but they were never used and it was essentially moral armament. By the spring of 1864 the Navy Department had received proposals for its tender for a new torpedo and torpedo boat and it accepted those of two engineers, Wood and Lay. The boat was a straightforward steam picket boat – 16·5m × 3m × 1·0m (54ft × 9ft 9in × 3ft 4in) – with armoured bulwarks. It carried a Dahlgren howitzer on the foredeck and mounted in guides on its starboard side an 11m

One of the handful of David submersible torpedo boats built by the Confederates during the American Civil War, seen here in Charleston Harbour, South Carolina, in 1865. The proboscis at her bow is the spar at the end of which a torpedo is mounted.

(36ft) pole to which a stubby copper cylinder, coned at one end, could be attached. This case contained the explosive and when the pole was run out some 5·5m (18ft) ahead of the launch and about 3·1m (10ft) beneath the surface, the cylinder could be hauled free of the spar; an air chamber was built on to the top half of the case so that when the firing line was tugged the charge would expend itself upwards to the enemy's hull. These ideas represented a considerable advance in the thinking of torpedo boat warfare, eliminating the necessity for a spar torpedo boat to make contact and providing the opportunity for accuracy and remote firing control. The merits of the Wood and Lay boat were successfully utilised by the unstoppably heroic William Barker Cushing, a lieutenant from Wisconsin, who volunteered to sink the rebel ironclad *Albemarle,* which was bullying its way along the Roanoke River. He approached *Albemarle* on the night of 27 October 1864, and at full speed ploughed his boat (usually known as Cushing's Launch No 1) on to a pen of logs surrounding the ironclad. He ran out the spar and released and fired the torpedo while 4·6m (15ft) from *Albemarle,* which was holed by the explosion and sank. The launch also sank, probably as a result of the turbulence of the explosion.

Cushing was the only man to escape from the wreckage, thirteen others were either drowned or captured, and he was promoted and given a vote of thanks by Congress for his valour in the action. In 1890 his name was given to the first American sea-going torpedo boat. Other Federal vessels were fitted with Wood and Lay torpedoes but were never in action. After the success of their launch Wood and Lay designed and built a heavily armoured 25·6m (84ft) torpedo boat, the *Spuyten Duyvil,* which disgorged a spar torpedo beneath the surface from a clever but intricate mechanism which occupied most of the forward compartment of the boat. She was also partly submersible and when she flooded her ballast tanks became 49·5cm (19·5in) deeper in the water, with a new draught of 2·8m (9ft 1in): she was not a fast boat. Nor did she have the opportunity to show off her weapon in action but just served as a picket launch in the James River. After the war, though, her ability to retract the spar and reload with another torpedo was useful when she was engaged in blowing up Confederate mines and obstructions.

In all the only ship sunk during the Civil War by a torpedo boat was the Confederate *Albemarle.* The South damaged Federal vessels on only two occasions with torpedo boats *(Housatonic* being sunk by a submersible) though

22 vessels were destroyed and 6 damaged by various explosive devices. Their value to the South was not as great as was believed by European navies in the years immediately after the war. The mines were the most damaging because they made the North's blockade more difficult to enforce, but as this had the effect of prolonging the war it was an equivocal weapon. Nevertheless, major navies started to think about armouring their ships beneath the waterline as well as above and many arguments were wasted and many experiments were made to make fleets immune to the terrible torpedo – not that many of them were actually attacked by a spar torpedo, although the Russians' successes against the Turks in 1877 and 1888 and the French against the Chinese in 1884 compounded the growing fear of the weapon. The mine, turned into an efficient, reliable weapon by the German, Dr Hertz in 1868, caused consternation, though it was only used defensively in the Austro-Prussian War of 1866 and the Franco-Prussian War of 1870.

As early as 1867 an Admiralty constructor, Nathaniel Barnaby, realised that:

> 'however thick the armour, the bottom of the ship was left as weak and defenceless as ever...In view of these new and undeveloped, but terrible agencies [torpedo boats], it was desirable not to create a large fleet like that recommended, at the cost of many millions of money, which, although secure from gunpowder above water, might easily be rendered obsolete and useless, because undefended from gunpowder fired from beneath.'[8]

E. J. Reed, the chief constructor of the Royal Navy, was very conscious of the power of underwater explosions and realised that 'No ship's bottom can, in fact, be made strong enough to resist the shock of such an explosion; and the question consequently arises, how best can the structure be made to give safety against [this] mode of attack'.[9] The solution was the introduction of double bottoms (first used by Brunel as insurance against grounding and poor navigation in *Great Eastern*) and watertight compartments below the waterline. This practice became widespread during the rest of the nineteenth century but there was little offensive mine warfare until the Russo-Japanese war of 1904, when Russia lost six warships and Japan ten to mines. Thus it is possible that the ingenuity of Reed's naval architecture was misapplied, but it is indicative of the grip that mine and torpedo warfare held on naval thinking. Among other defences proposed was a pemmican of cork shavings, chalk and sodium silicate fastened all round a hull in huge compartments of sheet iron. This was suggested in 1872 by the chief engineer of the Austrian navy, J. N. Moerath. In the same year C. W. Merrifield suggested a triple skin hull and cellular subdivisions, but it was impossible to subdivide effectively the boiler and engine rooms of warships, which could easily occupy half the length of any hull. Plans were also put forward for 'torpedo catchers', Heath Robinson devices which snipped mooring lines or scooped mines into baskets. Nets were also proposed, but at this stage were rejected by the Royal Navy as an ineffective anti-torpedo defence. There was a vociferous school of thought which considered that, as the torpedo was invincible, the Navy should trust not in armour plating but build faster ships, to outrun torpedo-carrying boats. As John Scott Russell said, 'The best mode of encountering gun-cotton torpedoes is to keep away from them. I think we had better ... study first how we are to carry and manage torpedoes ourselves, and then how we are to keep out of the way of other people's torpedoes.'[10]

In the mid-1870s the Admiralty was indecisive about what course to take, and there was almost an appeal for a Royal Commission, but by 1876 Barnaby was Director of Naval Construction and had formulated an attitude to torpedo warfare which was acceptable to the Royal Navy.

'The assailants ought to be brought to bay before they could get within striking distance of the iron-clad by consorts armed like the attacking vessels, with the ram and torpedo, which may take, like them, the chance of being sunk. The defence against the torpedo must be sought for, not in the construction of the ship alone or mainly, but also and chiefly in the proper grouping of the forces at the points of attack. Each iron-clad ought to be a division defended against the torpedo and the ram by smaller numerous but less important parts of the general forces. If the foregoing considerations are correct, there is still a place in naval warfare for costly iron-clads with thick armour and powerful guns. There is a place also for association with them of unarmoured vessels, armed with the torpedo and manned by brave men.'[11]

The influence of Barnaby's thinking was such that the Admiralty decided that a pioneer torpedo boat was necessary, particularly in view of the spar torpedo launches English yards were building for other countries.

The problem which faced the torpedo enthusiasts was how to strike the enemy without incurring his wrath. The spar was all very well, and was actually fitted to numerous Royal Navy launches and pinnaces, but even an enterprising lieutenant might be forgiven for wishing for a device less liable to sink his own boat as well as that of his protagonist. In 1869 there was one plucky attempt to create a better weapon than the spar torpedo. It was the sea torpedo (a mine in reality), designed by Captain John Harvey RN and his brother Commander Frederick Harvey RN. It consisted of a rhomboidal iron-bound elm case 1·5m long × 15·6cm wide × 0·5m deep (5ft × 6⅛in × 1ft 8¾in); there was a smaller model at 1·1m × 12·7cm × 0·4m (3ft 8in × 5in × 1ft 6in) with an interior case of copper. When towed through the water, at not less than 6 knots, on a 137m (150yd) cable, it travelled at an angle of about 45 degrees to the ship's course, like the otter of a trawler. It was this feature which enabled a skilful operator to bring it into contact with an enemy hull, where it was triggered off by a large projecting lever. Compared with the dangers of spar torpedoes, this invention gave the attacker total immunity from being destroyed by its own sting, but it was a singularly cumbersome weapon to use. It required at least three operators on deck and close contact with the wheelhouse, and it was virtually impossible to use at night. Nor could it be installed on a vessel much smaller than a sloop but when a vessel had the deck space and men to spare, it could have been fitted as an auxiliary arm for use at close quarters. The Navy toyed with the Harveys' torpedo for a while but it was not tested on active service and was soon eclipsed by the Whitehead.

The idea of the self-propelling torpedo came to an Austrian marine artillery officer in about 1860. His plans showed a small propeller-driven boat powered by a steam or a hot-air engine, packed with guncotton in its bow, which could be sent off like a model boat but with long lines fastened to a tiller quadrant for steering. This man died before he could build his invention, but the plans passed to a captain in the Austrian navy, *Fregattenkapitän* Giovanni de Luppis. He built a clockwork-powered model, loosely based on the plans, and presented *Der Küstenbrander* (coastal fireship) to the naval authorities. Its potential was realised but the project was considered unworkable and in need of more development. At this time one of Austria's foremost marine engineers was the Englishman Robert Whitehead (1823-1905), manager of Stabilimento Tecnico Fiumano, of Fiume near Trieste.

He had designed and built the engines for the ironclad battleship *Ferdinand Max,* in 1865 and was highly regarded for his excellent engineering skills. In 1864 de Luppis went to see Whitehead, who was attracted by the idea, but the two men could do little to improve the model and the partnership was concluded. Whitehead continued to think about the idea, realising that the weapon had to be independent of control-lines if it was to be of interest to the Austrian navy. He spent two years working on the project, designing a compressed-air engine and a guidance system to produce a working prototype which ran beneath the surface and carried 8·2kg (18lb) of dynamite to its target. It was an astonishing achievement and in 1866 the torpedo was demonstrated to the Austrian navy. It was fired below the surface from the bow of the gunboat *Genese.* Its target was the *Fantasie,* a 70m (200ft) yacht moored 640m (700yd) from *Genese* and protected by a net. Of the first 54 shots only 8 hit the net, because there was no device enabling the torpedo to maintain an even depth. This 3·5m (11ft 7in) torpedo, weighing 157·1kg (346lb), travelled at 5·7 knots. In three weeks Whitehead had designed and incorporated a balance chamber for level running and increased the speed to 6·8 knots. At the second trials fifteen out of thirty shots struck the net. When the full-sized 4·3m (14ft 1in) version was tested it was fired from a tube 4·3m (14ft) beneath the surface and was reasonably accurate. The Austrian torpedo commission considered that the certainty of hitting was as great as could be expected from such an arm, and recommended that the government should purchase the 'secret', as it was known. For an imperfect weapon it was expensive. According to Lieutenant F. M. Barber of the United States Navy[12], by 1872 Whitehead had spent £40,000 developing the torpedo, so it is little wonder that though the Austrians adopted the Whitehead they had to forego the option of exclusive manufacturing rights.Whitehead looked around for a buyer. The American navy was interested in 1869, but jibbed at the price of £20,000 preferring a home-grown automobile torpedo provided for the US Navy in 1870 by Captain J. A. Howell. News of Whitehead's torpedo first reached the Admiralty in January 1867, following a Foreign Office report about the trials held at Fiume in December 1866. The Director of Naval Ordnance, Rear-Admiral Cooper Key, was sceptical of the weapon's abilities and nothing happened until he received another report, in September 1868, from Vice-Admiral Lord Clarence Paget, Commander-in-Chief in the Mediterranean. Paget suggested some action might be taken, and on 17 September Key recommended a trial of the Whitehead.

Until this time what the Royal Navy understood by a torpedo, was actually a mine. Their only use was for harbour defence which, up to 1866 had chiefly been the responsibility of the Royal Engineers. In the summer of 1867 a course of mine instruction was started at the Royal Navy's gunnery school in Portsmouth, HMS *Excellent*. Experiments with Harveys' sea torpedo followed and the branch so expanded that in 1872 HMS *Vernon* became attached to *Excellent* and four years later was commissioned as a separate establishment. This was largely due to Commander John Arbuthnot Fisher, who ran the mine courses in *Excellent* from 1872, and who is largely credited with the introduction of torpedoes into the Royal Navy. However, it must be understood that Fisher's enthusiasm was for electrically controlled observation mines, and also Harveys' sea torpedo. Only the third edition of his three reports[13] carries any favourable comment about the Whitehead fish torpedo. (The term fish was invariably used in referring to the Whitehead to distinguish it from stationary 'torpedoes'.)

Following Key's approval, three officers from the Mediterranean Fleet were appointed by the Fleet Commander-in-Chief, Vice-Admiral Sir Alexander Milne, to report on the Whitehead trials at Fiume. The report[14] was decidedly favourable, but the new Director of Naval Ordnance, Captain Arthur Hood, was concerned about the torpedo's ability to be fired at moving targets through rough water. Hood's opinions may well have been influenced by Fisher's, and so Whitehead was invited to bring his weapon to England for further trials. These were conducted in the Medway off Sheerness in 1870 by a committee comprising Captain William Arthur, Captain Morgan Singer, and Lieutenant A. K. Wilson, with the result that the Whitehead torpedo was adopted by the Royal Navy in April 1871.

As England's expert on mines, Fisher had been in close touch with German developments and there was a continual exchange of information between HMS *Vernon* and the Prussian Torpedo Department at Kiel and Wilhelmshaven. As a consequence the Admiralty rejected the design of the Hertz contact mine while Messrs Schwarzkopf of Berlin were enabled to produce their own Whitehead, with a phosphor-bronze casing instead of steel. Nevertheless, the talents of Whitehead benefited England. The government bought the manufacturing rights for £15,000 and contracted Whitehead to instruct officers at *Vernon* in the 'secret' of the torpedo. This was the ingenious depth-keeping device, in which a pendulum was activated by the pressure of water to maintain a set depth when in motion. The torpedoes were manufactured at Woolwich Arsenal but the Fiume factory also built torpedoes for England and other countries.

It was this weapon, the Whitehead, which became the catalyst for the torpedo boat, the most precocious naval weapon of all time; within ten years of its introduction no self-respecting navy was complete without its pack of Whitehead torpedo boats. Beleaguered nations, poor nations, peaceful nations, even nations without a seaboard, all wanted – and were able to afford – divisions and even fleets of torpedo boats. With the torpedo any half-pint navy became a threat and could bait senior powers. It was only the top sea-dog, Britain, who was slow to equip herself with hordes of torpedo boats but this was not too vexing as most of the world's torpedo boats were built on the banks of the Thames, and Messrs Thornycroft and Yarrow were coining it for *pax Britannica.*

2 TORPEDO BOAT

The search for speed under power was continually thwarted by the dogma of the wave-line theory, formulated by William Froude after 1868, and vigorously supported by John Scott Russell, who maintained that the faster a ship travelled so the resistance it encountered proportionately increased, and more speed could only be achieved through greater length. As steam muscled in on sail during the mid-nineteenth century, bigger and bigger iron vessels were built, but the speeds of liners and warships rarely topped 15 knots. Scott Russell had the correct answer but the wrong equation; the hull of a displacement boat has a theoretical maximum speed (reckoned by the square root of the length of waterline) but by adjusting the design parameters, weight, power, length, and beam, and their relationship to each other, startling gains in speed could be made. The developments were twofold: revolutionary, and those made possible by the assiduous theoretical and practical research of boatbuilding and engineering pioneers. The first naval steamboat was a converted 12·8m (42ft) launch fitted with two 2-cylinder engines in 1863 by the chief engineer of Portsmouth dockyard, Andrew Murray. This boat, *Experiment*, attained 6½ knots. A few more 12·8m (42ft) pulling launches were fitted with ponderous 18·7kW (25ihp) steam engines after 1864, but they were not satisfactory.

In 1865 designs were invited from the engineering companies of Maudslay, Penn, and Rennie, the latter abandoning the conventional locomotive boiler for a multi-tubular boiler with two surface-condensing 32·1kW (43ihp) engines. Rennie's machinery weighed 3·7 tonnes (3 tons 13 cwt) and, fitted in Steam Launch No 10, achieved 7·7 knots in 1866, and by 1868 34·3kW (46ihp) launches were reaching 8·4 knots. This twin-screw boat came into general use as harbour launch, admiral's barge, and tender. More were fitted with spar torpedo equipment but coal consumption was quite high at 2·3-2·7kg (5-6lb) per indicated horse-power per hour, which limited their steaming duration. The standard for steam launches was set by the Cowes boatyard of White's, whose Lamb and White lifeboats were acquiring a world-wide reputation. In 1861 the yard supplied an 8·2m (27ft) steam cutter to HMS *Sylvia,* a gunboat on surveying duties. In service conditions the cutter attained 8¾ knots and was a considerable success. Her designer, John Samuel White (1835-1915) then met the Birmingham engineer George Belliss and in 1867 they produced a single-engine 11m (36ft) pinnace which achieved 8 knots with much greater engine efficiency and economy than the Admiralty 12·8m (42ft) launch. This boat was not taken into service and White pressed on with his lifeboats until

1880 when an order was placed for a 14·6m (48ft) pinnace. The pinnace was introduced into service after the steam launch had been adopted as part of warships' boat complement. She was of finer form than the launch, usually single screw and could be built in 9·1m, 11·3m, 12·8m, and 13·7m (30ft, 37ft, 42ft, and 45ft) forms with compound engines and forced-draught boilers. With a single 64·8kW (87ihp) engine this boat ran at 12·15 knots while a 90·3kW (121ihp) twin-screw boat achieved 13·4 knots. In confined waters speed was not the only requisite and White dramatically increased the manoeuvrability of his pinnace, standardised to 12·8m (42ft) in 1881, by fitting a rudder astern of each propeller and cutting back the keel deadwood to about two-thirds of its conventional length.

This speculative boat was bought into the Royal Navy and attached to HMS *Inflexible* where its handiness commended it to Captain John Fisher. Dubbed a turnabout, it was claimed she could turn 360 degrees in thirty seconds. Properly handled, most boats can be turned around in their own length but, with the cumbersome engine-reversing procedure of the 1870s and 1880s, White's boat undoubtedly represented an advance in forward-speed manoeuvrability. He applied the turnabout design to another steam boat in 1883, which became the Royal Navy's standard picket boat for many years. The boats were 17·1m (56ft) in length × 3·1m (10ft) beam, built with triple skin mahogany. White was the first boatbuilder to exploit fully the advantages – lightweight and firmness of construction – of planking boats with two or three thin layers of wood, layed diagonally across each other. If damaged the hull could be difficult to repair and the sheets of calico between each skin sometimes caused the timber to rot by holding water between the skins, but where homogeneous strength was required double diagonal planking was not surpassed until the introduction of glass fibre. Fitted with 111·9kW (150ihp) engines they made 14½ knots and were the fastest craft of their kind in the Navy and probably the most seaworthy. Many were fitted with spar torpedoes and side-dropping gear for Whiteheads as they became available; in such form they constituted second-class torpedo boats. Handling them called for special skills and torpedo boat coxswains received 6d a day extra pay.

The advantages in design, construction, weapon, and machinery reliability between 1860 and 1890, wrought from the War Office's increasing desire for harbour and coastal defence and the ability of companies such as J. Samuel White to provide suitable vessels, meant that there were some 165 steam launches and pinnaces employed in coastal defence work by 1885. The irony of this development was that the main threat was considered to be torpedo boats, and it was England that launched the fashion for the glamorous and rakish little hell-raisers.

Two men, Alfred Fernandez Yarrow (1842 – 1932) and John Isaac Thornycroft (1843 – 1928) were responsible for introducing fast steam launches into England and then meeting the huge demand for launches for pleasure and naval requirements. Yarrow, and his partner James Hilditch, were precocious inventors and with the money earned from their steam ploughs Yarrow was able to buy the Thames yard of Folly Wall at Poplar on the Isle of Dogs. In 1868 Yarrow received his first commission for a steam pleasure-launch. It was a 7·3m (24ft) vessel with a cabin for four people and a perch for the engineer. The launch was a great success and, though Yarrow actually lost money on the sale, he put a photograph of the boat in every river-side tavern between Oxford and Poplar. This campaign brought in many more orders and in the next seven years the yard produced about 300 launches. In 1872 Yarrow fitted out a 9·1m (30ft) launch as a spar torpedo boat. He was probably the first private builder

The pioneering triumvirate of high-speed steamboat design and construction: Alfred Yarrow, Jacques-Augustin Normand, John Thornycroft.

to do so but his attention was diverted from the development by various orders from explorers and missionaries for shallow-draught steamers to probe the waterways of Africa and South America.

In addition to his natural skills – he had built the 11m (36ft) *Nautilus* by the age of nineteen – Thornycroft was trained in the arts of engineering and shipbuilding. The year after *Nautilus* was launched he completed the 12·2m (40ft) *Ariel,* and her very light construction and speed of 12·2 knots revealed Thornycroft's preoccupation with fast light hulls. After building five launches he acquired Church Wharf at Chiswick in 1866, and from there he created vessels that, with Yarrow's, were the fastest in the world for the rest of the century. The boat which brought Thornycroft's skills to public notice was the 15·2m (50ft) steam yacht *Miranda,* launched in 1871. She was 13·9m waterline length × 2·0m × 0·8m (45ft 6in × 6ft 6in × 2ft 6in) and had a two-bladed screw 0·7m (2ft 6½in) in diameter. Maximum engine speed was 600rpm and she achieved the startling speed of 16·2 knots, with a two-cylinder engine of 47·7kW (64ihp). To credit this achievement for public satisfaction it was neccessary for an independent consulting engineer, Sir Frederick Bramwell, to reveal that no trickery was involved.[1] Thornycroft had simply burst through professional convictions about a water-speed barrier with a light easily driven hull, powered with light machinery. On hearing Bramwell's report, Scott Russell made the point that he had tried to accomplish such speeds as early as 1840 with experimental hull forms, but of course suitably lightweight steam engines were not available then. The fortune of the torpedo launch and boat builders of the 1870s was that high-speed boilers and engines were either becoming available or were being produced by the boatbuilding engineers themselves. The only tangible benefits of torpedo boat development were the rapid testing, rejection, or adoption of numerous innovations in construction techniques and boiler, propeller, machinery and hull design.

However, the craving for speed was exercising other minds than those of the engineers, notably that of the Rev Charles Meade Ramus. From Trinity College, Cambridge, he acquired the living of Playden, near Rye, Sussex, where he experimented with home-made models of extraordinary novelty. His originality can be measured when the designs are seen against the centuries-old habit of the displacement hull, in which the rounded underwater body sits in the water and encounters resistance proportionate to the area of its wetted surface. In 1874 Ramus wrote: 'The idea that speed may be gained by so shaping a vessel as to obtain a reduction of resistance by lifting it on the water is by no means new.'[2] He could have been referring to the experiments of Scott Russell and Abraham Morrison of Pennsylvania, who patented a planing-hull design in 1837. 'It is a well known fact', Ramus continued, 'that flat-bottomed vessels, and all vessels whose bottoms approach flatness, when urged forward by a strong propelling force have their bows lifted, and when their bottoms are thus placed in a position inclined to the surface of the water some advantage in speed is gained.' A considerable advantage in speed is gained but, for want of suitable prime movers, the flat-bottomed planing-hull had to wait thirty years before it revolutionised high-speed boating. Ramus considered the instability problem of a hull built in one plane and produced the dazzling idea of building a hull in two parallel planes so that it might skim over the water on an even keel, balanced on its two areas of contact with the surface. In April 1872 he made a small model to satisfy himself that the idea was valid and, convinced, he wrote to the Admiralty. He did not want a grant or aid for his bisphenic hull form, just credit for the invention which he foresaw leading to water transport at speeds of '30, 40, or 50 miles an hour [26, 34·7, or 43·4 knots]'.[3] The means to drive a vessel at these speeds was absent but in another letter to the Admiralty in December 1872 he advocated two pairs of propellers, one set driving from the aft end of each plane. In 1873 he fitted two models of solid fir, one 1·1m × 14·6cm (3ft 9in × 5¾in) and the other 0·7m × 12·1cm (2ft 5½in × 4¾in), with rockets and achieved astonishing performances from which he derived his speeds for full-size vessels; but he misapplied Froude's theory of comparison between models and ships. With 170·1 grammes (6oz) of powder the rockets were soon extinguished, but overcoming its inertia the 3·2kg (7lb) model still travelled 60·4m (66yd) in 6 seconds. The 1·4kg (3lb) model ran for 96·0m (105yd) in 3 seconds, an average of 63 knots. His scheme was for ships of up to 2032 tonnes (2000 tons) to be borne across the oceans at similar speeds but he failed to realise that insofar as his models were severely overpowered so the amount of power necessary to raise his polysphenic ship on to a planing position was unattainable. He also suggested that rockets should be fitted on full-size vessels which, when packed with guncotton, could 'be used for destroying from a distance the ships and fleets of the enemy'. The Admiralty instigated tests on Ramus's project under Froude's guidance at his new tank-testing and research establishment at Torquay, but Froude reported adversely on Ramus's claims. Ramus then submitted an outline for a polysphenic ship which skimmed on three inclined planes but Froude was again discouraging. Ramus claimed that Froude used inferior rockets but Froude, one of the greatest naval architects of his day and rightly regarded as an oracle by the Admiralty, claimed that he simply used the same shop-bought rockets used by Ramus. The fault lay with the Admiralty, who inadequately briefed Froude, though he might have shown a little more initiative in developing the idea. Later Froude did 'admit that it had seemed to me certain that at *some* assignable speed the skimming action would become so perfect as to obliterate, or virtually obliterate, water resistance'.[4] Despite Ramus's two pamphlets and petition to Parliament the project was stifled by the squabble and the strongly enforced Explosives Act of 1875, which prohibited the testing of rockets, handicapped his enthusiasm.

Ramus's achievement was to free a hull from the restrictions of travelling through the water, where additional power cannot overcome the limitations of the hull to make the vessel travel faster. When a craft is able to travel over the water its resistance is reduced and it requires correspondingly less power; the limits to high-speed travel over water are the glutinous nature of water itself and the necessity for a vessel to be able to perform its function when travelling at speed, whether for pleasure or warfare.

The Ramus affair had fizzled out by 1877 when John Thornycroft lodged a patent for 'An Improved Method of Reducing the Friction of Vessels when Travelling on the Water'.[5] There is nothing to suggest that Thornycroft colluded with Ramus in achieving skimming models of the same principle. Though Sir Nathaniel Barnaby (appointed Director of Naval Construction in 1875) showed interest in Ramus's proposals, his son Sydney Barnaby did not join Thornycroft's company as a naval architect until about fifteen years later and Thornycroft had no contact with the Admiralty until 1876. Moreover Thornycroft built his first model, a spherical design about 1·5m (5ft) wide, in 1873. In his patent specification he outlined the elements both of the hydroplane (which is what Ramus's work amounted to, though he made no attempt to patent his discovery) and the hovercraft, the latter seventy-three years before its fanfared invention by Sir Christopher Cockerell. Thornycroft was well aware of the necessity of keeping the hull on top of the water, but he underestimated the power of the hydroplane and air-cushion principles to perform the function individually, and he combined both aspects:

'It has often been proposed to make a vessel rise to near the surface of the water by giving an extended inclined bottom and high speed, but the friction of the surface has so far as I know always rendered the position impracticable owing to the magnitude of the force required to overcome the friction. Now, my improvement is designed to provide an extended supporting surface almost without friction with a view to the attainment of quick speeds. In order to make a vessel which should rise to the surface at a moderate speed a large amount of water should be acted on vertically and a downward motion impressed; the width of the vessel measures the surface acted on in a given distance, so the width of the vessel should be as great as practicable.'

This accounts for the unlikely near-spheroid form of the model Thornycroft is discussing, which had a step, or inclined plane as Ramus would have said, describing a curve across its entire width. Later experiments showed that the step need be no wider than the beam of a conventionally shaped vessel. However, to bolster the support of the step, and this was doubtless the 'improvement' (perhaps over Ramus's ideas), air was pumped through a blow-hole in the centre of the model just aft of the step, to elevate the hull and rid it of the bogey of surface friction. This idea was further developed in later models of conventional hull shape, in which the compressed air was more easily contained: 'In order to reduce friction of a vessel when travelling over water I interpose a layer or body of air between the bottom of the vessel and the surface of the water, which air I confine within a cavity of the bottom of the vessel.' When working out designs for the Admiralty's first torpedo boat, Thornycroft built an alternative model which embodied a concave bottom, but he rejected the design, probably on the grounds of structural weakness and the fact that its resistance at low speeds would have been unacceptably high.

Following on the fame of *Miranda*, Thornycroft built another six fast launches in 1872-3, the biggest, *Sir Arthur Cotton*, astounding London with her unprecedented speed. She was a 26·5m (87ft) vessel designed for service on Indian waterways, and was powered by a 2-cylinder non-condensing 246·2kW (330ihp) engine. Her trial speed was 21·4 knots, and though this was largely due to lightweight construction she was undoubtedly the fastest boat in the world. It was probably news of this achievement which led the Norwegian government to order from Thornycroft in 1873 the first purpose-built torpedo vessel. She was a 17·4m × 2·3m × 0·9m (57ft × 7ft 6in ×3ft) launch with a 67·1kW (90ihp) compound engine and could steam at 14½ knots. *Rapp* was constructed in steel plates, divided into six watertight compartments and had two steel hatch covers which could be fitted over fore and aft wells when the vessel was under fire. The helmsman was encased in a small dome with holes punched around it at eye level. Amidships the captain could direct operations from a hood which had a 6mm (¼in) slit all round it. Delivered in 1874, she carried a 4m × 22·9cm (13ft × 9in diameter) towing torpedo in davits. This represented a curious sidestep in the development of torpedo boat armament by using the principle of Harveys' torpedo, though the shell-like nature of the hull and decks made it impracticable to fit and operate a spar torpedo. When released from its davits the weapon was towed, by a line attached to the funnel, at 40 degrees from the boat's course, and thus into the hull of the ship under attack. Unlike France, Germany and Italy, Norway had not bought manufacturing rights for Whitehead's torpedo, so it was a natural step to utilise the mechanics of Harveys' well-publicised system.

In the same year Thornycroft built a 16·6m (54ft 6in) launch in steel, *Choutka,* which was bought by the Russian navy. The Russians fitted a spar torpedo to the 16·9-knot boat and together with a handful of other torpedo craft she served in the war against Turkey in 1877. Later in 1874 the yard completed two other 17·4m (57ft) torpedo launches, one each for the Swedish and Danish navies, while Austria received a 20·4m (67ft) 18·3-knot torpedo boat rigged with two torpedo spars. In 1875 two of the 20·4m × 2·6m × 1·3m (67ft × 8ft 6in × 4ft 3in) boats were delivered to France for naval evaluation (and two years later the Havre shipbuilders Augustin-Normand produced the first of their many successful torpedo boats). This larger size of boat was built with heavier plates than the 57-footers and the armour plating extended down to the waterline. Engines of 149·2kW (200ihp) were fitted to cope with the extra weight and the armament was a spar torpedo rigged port and starboard. Two 23·2m (76ft) types were constructed in 1876-7, one each for the Dutch and Italian navies. The 186·5kW (250ihp) Dutch boat carried conventional outrigger, or spar, torpedoes, but the Italian boat *Veloce,* 18·27 knots, had a 238·7kW (320ihp) engine and was reputedly fitted with a Whitehead torpedo, the first torpedo boat in the world to be so armed. The importance of working the boats at sea was coming to be realised; instead of the flat-decked low freeboard design the 76-footers were given more freeboard forward, which made them drier in a seaway. There was some trepidation about the strength of these lightweight craft but all the plates were tempered and all the hull plates had some curve in them, and with extensive framing and machinery braced to the hull they were flexible and strong craft. Because of the speed advantage these early torpedo launches had their propellers mounted well clear of the hull and even astern of the rudders (modern powerboats also fit propellers well astern of the hull) but this practice was discontinued as it came to be realised that the manoeuvrability resulting from rudders aft of propellers was more useful than the loss in speed of one knot or so. As if to dispel fears about the frailty of their craft, Captain Koren of the Norwegian torpedo boat sent a letter to Thornycroft's partner, John Donaldson, saying:

'When we got out of the above-mentioned fiord, it was blowing a stiff double-reefed topsail breeze, with a corresponding but confused sea . . . the wind was freshening, and the boat certainly did not *look* like a sea-going craft; but we had come thus far and I thought it best to push on. We had about fifteen minutes to run with the sea on the beam, and when we lost shelter from the rocks, her movements were quite extraordinary, alarmingly quick, and great. However, she shipped no water, and having observed her for some little time, I knew there was no danger. . . .We got to Horten in good time, and all safe, and after my opinion, the boat is much more sea-worthy than one could possibly think, judging from a hasty glance only.'[6]

Yarrow had also built some spar torpedo boats for foreign navies by 1876, and naval authorities and officers were aroused by this fashion to such an extent that the Admiralty eventually commissioned Thornycroft to build a torpedo boat for the Royal Navy. It was a surprising decision for a navy which was indisputably in control of the seas, for the torpedo boat at that time was a last-ditch weapon, pottering about on harbour defence and ready to force a way out of blockaded ports. The spar torpedo became quite efficient during the 1870s. Fitted with McEvoy's electric detonator which gave the operator the choice of impact or controlled firing, it was in use in France, Norway, Denmark, Austria, Sweden, Italy, Holland, and Russia, even before these navies had any substantial knowledge of its capabilities in warfare. They were mesmerised by the high speeds of the launches and convinced of the superiority of the torpedo by the American Civil War. Moreover, they were cheap: the Thornycroft 57-footers cost £1800 each, the 67-footer £3400, and the 76-foot type £4750 (all ex-torpedoes); the number of crew averaged between ten and fifteen. The commissioning of the Royal Navy's first torpedo boat could well have been politically motivated. During the age of the transference of sail to steam power,

France had frequently led Britain, pioneering the use of the propeller, explosive shells and armour cladding in capital ships and selecting a superior type of breech-loading gun. A French naval representative had even beaten the English delegation to Fiume in 1868 to investigate Whitehead's torpedo. Nothing came of that, but when Whitehead visited London that year he threatened to offer the torpedo to the French unless the Admiralty, which was far from dilatory in the matter, reached a decision. (The French almost got the invention for nothing when in 1870 Whitehead was apprehended in besieged Paris while travelling to Sheerness with two torpedoes in his baggage.)

The Admiralty capitulated to the torpedo boat lobby in 1876 and HMS *Lightning* was launched in 1877 and renamed *TB 1* in 1878. She was a 25·6m × 3·3m × 1·5m (24·7m on the waterline) (84ft × 10ft 10in × 5ft – 81ft) single screw boat, displacing 29·2 tonnes (28·7 tons). She had one locomotive-type boiler and a two-cylinder compound engine developing 339·43kW (455ihp) at 440rpm. On her preliminary trials on the Thames she reached 19·4 knots but with gear and torpedoes aboard her speed was 18·5 knots. The weight of the machinery, with water in the boilers, was about one-eighth the weight of ordinary commercial engines. This achievement was at the expense of coal consumption, which was between 70 per cent and 100 per cent more than usual. Thornycroft pointed out that this was no reflection on the engines, but on the fact that the boiler was proportionately overworked.[7] With bunkerage of 7·1 tonnes (7 tons), *Lightning* had a range of two or three hours' steaming at full speed, and so had a restricted radius of action. Even cruising at 11 knots or so, she could barely cross the Channel to Cherbourg from Portsmouth, let alone execute some speedy action against the French fleet. Her self-imposed rôle was defence work. The task of sinking enemy ships with spar torpedoes was given to 18·3m (60ft) second-class torpedo boats, a type developed from Thornycroft's 57-footer, which were carried aboard larger vessels and lowered into the sea when required for some stealthy cutting-out exploit. Nevertheless, *Lightning* was an outstanding boat and orders for a further eleven Lightnings were placed at the end of 1877.

In America the navy took advantage of James Herreshoff's coil boiler. Invented in 1873 and fitted in a boat in 1874, this was a 12·2m (40ft) launch, *Vision,* designed and built by the Herreshoff Manufacturing Company of Bristol, Rhode Island. The coil was a water-tube boiler, difficult to construct but very light, and it allowed steam to be raised from cold in about five minutes, far quicker than in locomotive boilers. There were a few torpedo boats in the US Navy in 1875, steam launches which had been converted to carry spar torpedoes, but in that year *Lightning* was ordered by the US Navy from Herreshoff with a coil boiler. She was a 17·7m (58ft) purpose-built spar-torpedo boat of 19 knots, decked fore and aft and designed by Nathaniel Herreshoff (1848-1938), responsible for many later fast steam and motor boats and six America's Cup defenders. After this Herreshoff built spar torpedo boats for various governments and in 1878 the company built a 18·0m × 2·3m (59ft × 7ft 6in) beam torpedo boat for the Royal Navy, which was keen to evaluate the coil boiler. She was a fully decked boat, with a steel deck and frames and steel plating above the waterline, timber below.

Like Thornycroft, Herreshoff's steam engines were superbly engineered and had lightweight moving parts, but the coil boiler was not acceptable to the Admiralty because once the tube had become scaled it had to be completely replaced. Herreshoff had stopped building these boilers by 1885, and in 1886 Thornycroft patented the superior three-drum type of water-tube boiler. Yarrow produced another version in 1889 and Normand in 1890.

America was not in the same state of martial tension as most of Europe, and had no need to defend her harbours or patrol coastal waters with torpedo boats. In 1887 the US Navy bought its second torpedo boat, *Stiletto,* a converted Herreshoff fast steam yacht of 1885. She was 28·7m (94ft) long and 3·5m (11ft 6in) in beam, and could cruise at 17 knots with a top speed of 23 knots. Also in 1887 the US Navy ordered from Herreshoff the 42·1m (138ft) steel vessel known as Seagoing Torpedo Boat No 1, commissioned as *Cushing.*

When Whitehead was working on the design for his locomotive torpedo he soon realised that to be of any use it would have to travel under water, not on the surface like Luppis's prototype, and all his firings at Fiume and Sheerness were from tubes mounted underwater, in emplacements or fitted in ships' bows. The Whitehead was naturally regarded by the Navy as another form of gun and the 264·2-tonne (260-ton) HMS *Vesuvius* was specifically designed and built as a silent battery from which to fire Whitehead torpedoes. Launched in 1874, her coke fumes were exhausted along the vessel's deck to evade detection and she carried ten torpedoes but could only steam at 9 knots. HMS *Polyphemus* was an ironclad built in 1878, with a 3·7m (12ft) ram as her primary weapon to protect the main battle fleet from torpedo boat attacks. But during her construction five torpedo tubes were added to her armament, against the wishes of her designer, Nathaniel Barnaby: four tubes were fired broadside, two above and two under the water, and a tube was fitted beneath the ram for line-ahead firing (which, with line-astern firing, was the usual practice on those foreign navies which carried Whiteheads). *Polyphemus* was not a success but she was the first vessel to be built as a direct antidote to the torpedo boat threat. Faster and more seaworthy vessels for this purpose followed six years later. Rather more usefully equipped was a large unarmoured frigate, HMS *Shah,* fitted with two torpedocarriages on her foredeck.

The Royal Navy first experimented with surface launching Whiteheads in 1875, when a 40·6cm (16in) torpedo was placed on a messroom table and shoved out of a port. The torpedo was seen to work and the Navy developed the alternative launching method, to the consternation of Robert Whitehead who protested that the 'delicate weapons are not meant to be fired like shot from a gun'.[8] *Shah* was the only Royal Navy ship to engage an armoured vessel before 1914, and the first to fire a Whitehead in action. On 29 May 1877 she and the corvette *Amethyst* clashed with the ironclad *Huascar* off Peru but neither gun nor torpedo gave much cause for celebration; *Shah's* armour-piercing shells failed to pierce and the torpedo, fired at 365m (400yd), was outdistanced by *Huascar,* which was manned by Peruvian revolutionaries. The 11-knot monitor retreated into the port of Ilo and after dusk *Shah* sent in some ship's boats armed with spar torpedoes and the longboat with a Whitehead lashed to its side. If the longboat failed to sink *Huascar* the outrigger boats were to carry on the attack. The rebel monitor fought off all attempts on her and steamed away over the shallows to hand back her illegal commission to the Peruvian navy the next day.

During the Russo-Turkish war of 1877-78 torpedo boats of the makeshift Russian navy were pitted against the Turkish Black Sea fleet, with results that compounded naval enthusiasm for torpedoes. The first Russian attack was launched on the night of 25 May 1877 by four spar-torpedo boats against the *Seife* in the Danube estuary. The leading torpedo boat, *Tsarevitch,* fired her outrigger under the Turk's stern quarter while a second boat exploded her torpedo amidships. Both torpedo boats were badly damaged, the first by rifle fire and the second, *Xenia,* fouled her propeller in wreckage while reversing from the sinking *Seife.* It became standard practice for Russian torpedo boats – and the vessels were designed with this hypothetical attack in mind – to charge to within 650m (700yd) of the target. Then the engines were shunted into reverse and, as soon as way was off the boat, the torpedo was fired and the boat retreated at full speed stern-first along the line of attack. This presented a smaller target to gunners on the attacked ship than a boat which curved away while steaming ahead. In a skirmish off Sulina on 10 June, Turkish boom defences foiled the second Russian attack and one Russian boat was sunk by shellfire. The Russians failed to use the spar torpedo to blow open the booms and thus clear the way for the boats using Harveys' torpedo, and at Nikopol thirteen days later two Russian boats again failed to penetrate the defences. There were also failures at the eastern end of the Black Sea, primarily attributable to the use of Harveys' torpedo. However, an attack was made with Whiteheads against the Turkish ironclad *Mahmoudieh* off Batum by *Tchesma,* with the torpedo slung beneath her keel on ropes, and *Sinope,* carrying the weapon in a tube against the side of her hull. Both torpedoes missed their target, at a range of 55m (60yd). In the second attack the Russians were either luckier with their individual torpedoes or had correctly set the temperamental guidance controls, for the same torpedo boats both scored hits on the 2000-ton steamer *Intikbah,* which sent her to the bottom.

Many of the Russian torpedo boats had been built to a design by Yarrow. In 1877 Russia placed orders for one hundred 22·9m × 3m beam (75ft × 10ft) boats, most of which were built in Russia and fitted with Yarrow machinery constructed in Germany. The 18-knot boats were assembled in the Baltic and many were transported by rail in a week from Leningrad (St Petersburg as it was then known) to Sevastopol for use against the Turks. British relations with Russia were strained at this time, with Russia pushing her frontiers towards Afghanistan and India, and in 1877 the Silurifico Whitehead company stopped selling its torpedoes to Russia. In the same year two Yarrow 21-knot spar-torpedo boats ordered by Russia were confiscated by the British government and added to the Royal Navy.

Thornycroft and Yarrow had considerable respect for each other's design and engineering skills but the relationship was tempered by their different successes, nearly all Yarrow's torpedo boat sales being overseas, while Thornycroft nobbled the home market, (during their heyday in the 1880s torpedo boats were commonly known in England just as Thornycrofts.) The principal feature of torpedo boats was their speed and the rivalry between the two Thames yards was influential in achieving high speeds through the water. Although the Royal Navy encouraged high speeds, it was tactically pointless considering the restricted range of the boats. Such torpedo boat actions as there had been showed that stealth, silence, and invisibility were the necessities for sinking enemy shipping. Naturally it was desirable to retreat quickly after firing the torpedo and to be able to weave away from machine-gun fire, but this discounted one of the arguments used to introduce torpedo boats – that they cost one-eightieth the price of an ironclad and, with their small crews, were to be reckoned as expendable. Moreover, they could not maintain their maximum speeds in a seaway, when they had to come down to 12 or even 9 knots. They were not designed for staying at sea and rolled furiously in rough seas, causing considerable discomfort on board. When a boat was handed over to the navy by the contractor it often took several weeks for the crew to achieve the performance vaunted for it, and there were many complaints from officers that their boats would not achieve their trials speeds. This was because the contractors had stokers and artificers skilled in the use of the boats' boilers and engines; the boats were run at their ideal trim and in weather conditions suited to the logging of the guaranteed minimum speeds. It is no wonder that a boat often fell in the expectations of its lieutenant and new crew. As one naval officer sourly remarked, 'A torpedo boat was a machine constructed to run a trial trip'. But the speeds kept rising. For its second batch of Lightnings the Admiralty ordered three boats from Yarrow, one of which, *TB 14* delivered in 1878, was the fastest boat in the world, achieving 21·9 knots. She displaced 33·5 tonnes (33 tons) and was powered by a Yarrow engine of 410·3kW (550ihp).

A White boat, *TB 19,* 28·5 tonnes (28 tons), achieved 21 knots. Between 1877 and 1880 twenty Lightnings were built for the Royal Navy but only nineteen were accepted; Rennie's *TB 16* was experimental and a failure. The last six boats each had two 35·6cm (14in) torpedo tubes. However all the boats were on the reserve list, periodically summoned for exercises and training. The Royal Navy maintained its supremacy with capital ships and there were no active rôles for torpedo boats. There was a considerable amount of bleating from the press and naval commentators over the paucity of Britain's torpedo boat fleet, and the growing numbers of French and Russian torpedo boats were proclaimed by some zealots as being model navies. The introduction of *Lightning* and her successors into the Royal Navy was an illogical step, ill-befitting the major naval power. Their only employment could be countering blockades on British ports – a hypothetical position in the 1870s – but the boot was put on an offensive foot in 1879 when *TB 1* was equipped with a tube on her foredeck for firing Whitehead torpedoes. Whiteheads were supplied to the Royal Navy's capital ships because of their destructive and deterrent characteristics, but a more practicable use was found for them by fitting them to torpedo boats. Though tube-discharging torpedoes had been fitted to two Italian Thornycrofts in 1878, and in the same year two French Thornycroft-built boats had a cradle on either side of the hull, it was not until the introduction of the larger and

The torpedo boat berthed inside the 2nd-class boat is *TB1* (HMS *Lightning*) at Portsmouth in the 1880s, the first Navy boat to be fitted with a tube mounting for launching Whitehead torpedoes. Within six years of their construction the first dozen torpedo boats in the Royal Navy were in a sorry state: 'they had been run at intervals training stokers and coxswains, or acting as vehicles for picnic parties and other useless purposes. When not actually in use they were allowed to rot and rust . . . they were hopelessly deficient in speed and general seaworthiness, the crews were totally untrained in their use, and loth to keep at sea in them for any length of time.' *vide* **Armstrong.** *National Maritime Museum, London*

faster Lightning–type boat, carrying tube-firing Whiteheads, that the combination became such a deadly threat. The deification of the torpedo increased and the rivals to British dominance, France and Russia, ordered even more torpedo boats, seeing in them the means to naval superiority at little cost. The Admiralty saw further into their limitations and only bought them as their size, and thus range and seaworthiness gradually increased. The Royal Navy's commitment to the torpedo as a substitute for gun and ram was proving ill-founded, for though it could be fired broadside from a capital ship under way, its effective range was about 365m (400yd). The armour-piercing gun was achieving greater range, albeit without commensurately greater accuracy, and by keeping their ships at shell range captains precluded the use of both ram and torpedo. Ram warfare between capital ships became obsolete and the torpedo of secondary importance, of use only in unusual circumstances.

Then a war in South America gave a boost to the torpedo bandwagon. Chile declared war against Peru and Bolivia in February 1879 over a disputed boundary in an area rich in nitrates. Bolivia did not have a navy but the navies of Peru and Chile were similar, and included a few torpedo boats. In April 1880 Chile imposed a blockade on Callao with two ironclads (one of which was *Huascar,* captured from the Peruvians), three support ships and two torpedo boats (a 15-knot American boat, *Guacalda,* and a 22-knot Yarrow boat fitted with three outrigger torpedoes, *Janequeo*). On 9 April *Huascar* escorted the torpedo boats into the heavily defended port but *Guacalda* stopped with engine trouble, lost the ironclad in fog, and only entered the harbour at 4·00am. She then collided with a fishing boat, breaking a torpedo spar, but her crew seized two fishermen, found out where the 1320·8-tonne (1300-ton) corvette, *Union,* was lying and closed her. *Guacalda* approached to within 13·7m (15yd) of the corvette but was brought up by booms. All the boat could do was fire her remaining torpedo against the booms and retreat, under fire from muskets and Gatlings. Both *Huascar* and *Janequeo* had missed the harbour because of the fog but the threat implied by *Guacalda*'s action was sufficient for the Peruvians to haul their ships into docks and so fence them in with booms that they almost became immobile. After bombarding the anchorage on 22 April the Chileans settled down to the weary business of enforcing the blockade. Here the two torpedo boats proved useful – ready to spot the 13-knot *Union* which was suspected to be planning escape. *Huascar* and *Blanco Encalada* both had electric searchlights, of little power, but which could have been useful in warning their own boats patrolling inshore of a raid by Peruvian torpedo boats. On the night of 23 April the Chileans made another torpedo attack and *Janequeo* closed with a similar sized armed launch but miscalculated the range and prematurely detonated her torpedo. In reply to this the Peruvians drifted mines out towards the blockade but they were spotted and destroyed. When patrolling the following month, the two torpedo boats encountered three guard boats and in the pursuit *Janequeo* closed with the 3048·0-tonne (3000-ton) ironclad *Independencia.* The torpedo boat was unable to use her bow torpedo, which was propelled over the stem by a small steam engine, and had to use one of the spars fitted beside the hull. Unable to clear the side of the ship, *Janequeo* fired her torpedo while almost alongside, tearing a hole in her own side. The *Independencia* sank at once but the pressures on *Janequeo's* watertight compartments proved too much, and she went down ten minutes later.

Torpedo boat activity was renewed by Chile six months later, when *Frezia,* a replacement Yarrow, and two Thornycroft outrigger launches, arrived at Valparaiso, where they were all equipped with one Hotchkiss machine gun. The allies' only move had been to scuttle one of their two American Herreshoff torpedo boats. On 6 December, *Guacalda* and *Frezia* spotted a Peruvian gunboat of about 30·5 tonnes (30 tons), *Arno,* and pursued her across the Bay of Callao towards the harbour. The battle was fought entirely with the Hotchkiss guns aboard each vessel as the torpedo boats were unable to close with *Arno* before she reached the shelter of the shore batteries. Both sides scored hits, but at too great a range to pierce the steel, with the exception of one shell which penetrated *Frezia*'s stern while she was retreating from the forts. The boat sank, drowning her engineer, but she was raised a few days later and was back in service by the end of the month. A machine gun such as the Gatling was little use on a torpedo boat because for every correction in training the gun had to be cranked into a new position; the Hotchkiss and Nordenfelt were free mounting and the gunner had immediate control to compensate for the lumpy passage of a torpedo boat under way. Their place in torpedo boats was questioned, but with this action, in which *Arno*'s own gunners were often driven from their posts by the Chilean fire, their importance as at least secondary armament was assured.

The last torpedo boat action of the war involved the Peruvians' remaining Herreshoff, which was taken fifteen miles along the coast from Callao and, on 3 January 1881, was seen leaving Ancon harbour after dark. She had two Lay floating torpedoes in tow but the following morning she returned, still hauling the torpedoes, having been unable to find the Chilean ironclads off Callao. Later that day the Peruvian boat was spotted by *Frezia* and a 1696·7-tonne (1670-ton) corvette, *O'Higgins,* which sank the Lay torpedoes and drove the Herreshoff boat ashore, where she was destroyed by the corvette's guns. After this the Peruvians sank the remnants of their navy and the war drew to a close.

The successful torpedo boat actions in this war were seized on by torpedo addicts and used to disseminate information about the invincibility of the torpedo. Of rather more practical use to the British government was the fact that the Whitehead was becoming more reliable and capable of greater range and speed. The Fiume 35·6cm (14in) Mark III of 1876 had the same range – 550m (600yd) – as the 35·6cm (14in) model tested at Sheerness in 1870, but it was three times as fast (21 knots) and carried three times the charge of guncotton, 26·3kg (58lb). After Whitehead sold manufacturing rights to Britain in 1871 he continued to develop his torpedo, and engineers at the Royal Laboratory in Woolwich Arsenal and later at the Royal Gun Factory, and those at the Fiume works, incorporated each other's improvements. The prototype engine, driven by compressed air at 26·0kg per square centimetre (370lb per square inch), was replaced by a 2-cylinder V-configuration in 1868, and in 1876 the 3-cylinder Brotherhood 29·8kW (40hp) radial engine was fitted, weighing only 15·9kg (35lb). Improvements in range went hand-in-hand with the capacity of compressed-air tanks to contain higher pressures and by 1882 bronze vessels were in use which stored air at 1500psi (105·5kgscm). In 1883 Froude advised a torpedo design committee that the sharply conical nose section, which had been accepted as automatically as pre-Sputnik schoolboys drew pointed rockets, was no faster than a blunt nosed section. Trials revealed that a blunt nose, which could carry more explosive, was 1 knot faster than the conical section and from 1883 Fiume and Royal Laboratory torpedoes assumed the familiar cigar-shape features. Between 1871, when volume-production began, and 1881, Whitehead sold 1456 torpedoes, including 203 to Germany, 218 to France, 250 to Russia, and 254 to Britain. Italy and Greece each had 70, Argentina and Belgium 40 each.

What concerned the Royal Navy more than the power of torpedo boats was the reverence with which they were held in France and Russia. In 1877

Britain had one torpedo boat, France 8, Russia 26 and Germany 1. By 1884 Britain had 19, France 50, Russia 115 and Germany 30. In the same years Italy had increased her boats from 2 to 18. The Royal Navy's policy was influenced by the possibility of war with France and the growing number of French torpedo boats was an impertinent threat to British naval supremacy. Instrumental in creating this threat was the ship and boatbuilder Jacques-Augustin Normand (1836-1906) of Havre, an engineer as talented as Yarrow or Thornycroft and with a strong family tradition of shipbuilding. His torpedo boats were a match for those of his English rivals and excited considerable attention in France. (Yarrow's sons were trained at the Havre yard.) To aid Britain's huge commitments overseas the Royal Navy bought many second-class torpedo boats, fitted with side-dropping torpedo cradles. These were 18·3m to 19·2m (60ft to 63ft) boats of 15 to 17 knots, which were carried aboard parent vessels in service on foreign stations. By 1885 the government gave in to the continuous growling for more torpedo boats and contracts were given to constructors for fifty-four first-class vessels. This was in spite of factors suggesting it was not altogether a wise decision and indicates the strength of the torpedo boat propaganda. The Royal Laboratory could only build 120 torpedoes annually (they cost an average of £300 each), and only 363 were aboard ships in 1884; 311 were needed for ships and boats which were ready for commission and another 320 for ships under construction and fitting out. The Fiume factory was largely supplying other navies and by its location was a risky source in event of war. Because of the shortage, the Admiralty was keen to establish another supplier and though Woolwich production was increased and extra orders placed with Fiume, the Leeds engineers, Greenwood and Batley, were given a contract for 100 torpedoes as well as drawings of the Whitehead, which was forbidden by the original contract of 1871.

At the same time Robert Whitehead offered to start manufacturing in

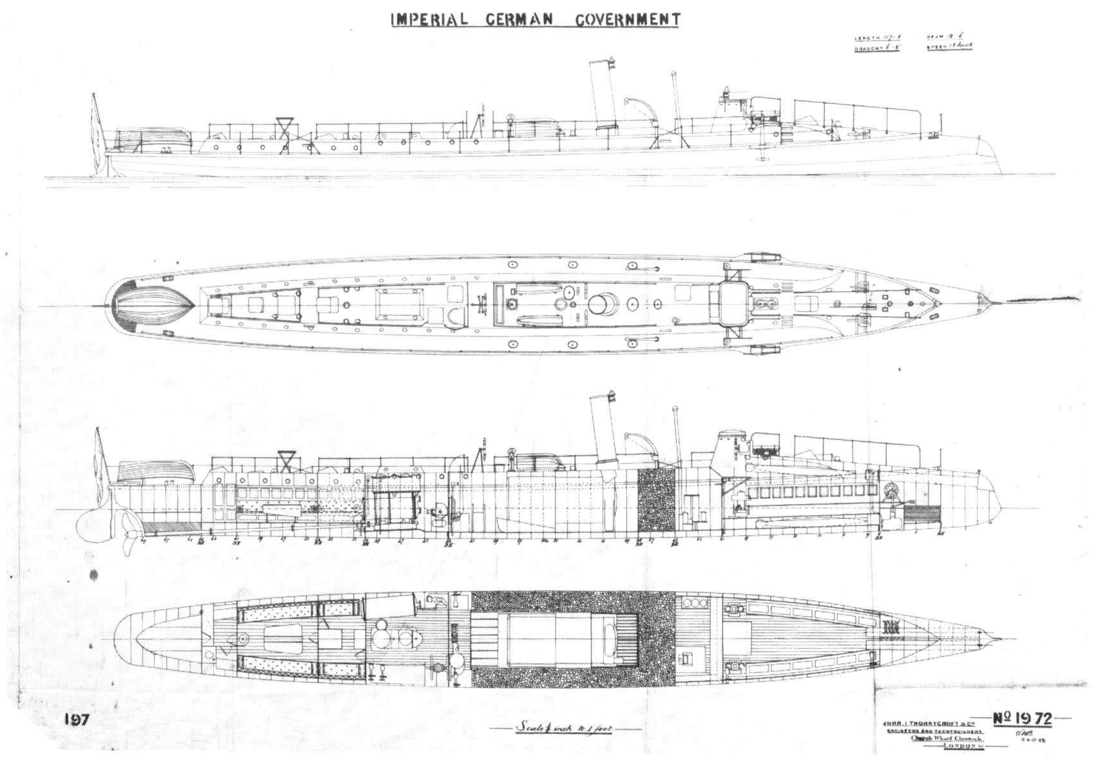

A 1st-class 19.9-knot torpedo boat built by Thornycroft in 1884 for the German navy. She was 35.9m (117ft 8in) long and was fitted with Whitehead's firing gear in her bow.

England, and though this was not taken up he opened a factory in Weymouth, Dorset, in 1890. John Thornycroft also wrote to the Admiralty, in October 1884, because it had been suggested to him by 'more than one foreign power that [Thornycroft] possessed exceptional facilities for the successful manufacture of torpedoes'[9]. Accepting that this was so, he proposed setting up an establishment at Hendon or beside the Thames estuary, but nothing came of the idea. The Admiralty acted more positively on a Royal Navy report of torpedo trials at Kiel in September 1884. The Schwarzkopf torpedo, suspiciously similar to Whitehead's, cost £100 more than the Whitehead and though it had less range, the Royal Navy's need was so urgent that a hundred were ordered.

Torpedo nets were advocated as sufficient defence against torpedoes, and a ruling during the 1891 naval manoeuvres was that ships with their nets out were held to be untouchable by torpedoes. To a considerable extent nets, made of steel mesh supported on 9·1m (30ft) booms, were impenetrable despite netcutters fitted on the nose of the torpedoes. They could not be used when a ship was under way but this was immaterial as it was considered that enemy torpedo boats would only attack by night in anchorages. (In a night exercise at Portsmouth in October 1879 HMS *Lightning* fouled her propeller in a ship's nets and was entangled from 9.00pm to 5.00am.) Electric searchlights and quick-firing guns were also considered excellent torpedo-boat repellents. In theory they were, though the searchlight could be shot out by the torpedo boat's own machine gun and it pinpointed the position of the attacked ship. From 1882 onwards the quick-firing Hotchkiss and Nordenfelt guns came to be seen as a panacea for the troublesome torpedo boat, firing twelve 2·7kg (6lb) shells per minute – but not by France and Russia, who kept on building, not losing sight of the fact that the boats were do-or-die and though six boats could be gunned out of action a seventh could sink a capital ship with one torpedo. Nordenfelt himself said:

> 'I believe machine guns have a good chance of destroying torpedo boats if the attack is made in full daylight and fair weather; but still there are many different occasions at sea – mist, bad weather, rain, darkness – when the torpedo boats would be able to attack favourably and would be able to see the hull of the big ship when [she] could not see them'.[10]

The biggest failing of the torpedo boats was their unseaworthiness. Highly stressed and insufficiently used, their menace was curtailed by breakdowns and poor seakeeping qualities. The hulls, like the yachts of the period, were narrow-beamed and they rolled mercilessly in any sort of a seaway. Some remarkable delivery voyages were made but at cruising speeds when men and machinery were under little stress. Yarrow's *Batum* was a 30·5m × 3·8m × 1·2m (100ft × 12ft 6in × 4ft), 22-knot torpedo boat built for Russia in 1880. She was the first sea-going torpedo boat and made a 4800-mile voyage from London to the Black Sea port of Nikolayev in eighteen days, an average speed of 11 knots. The success of this boat led to similar orders from six other navies, and the four boats for Argentina all crossed the Atlantic on their own bottoms, temporarily rigged as sailing boats. Ten years later two Yarrow 38·1m (125-footers), built in 1886, limped across the Atlantic to St John's with an escort, *Tyne*, that was herself pressed in the storms. One of the boat's officers recounted that they were 'battered and weather-beaten, their paint all gone, their stanchions bent, broken, twisted, and gone altogether; funnels caked with white salt, everything smashable smashed; very different were they from the trim boats which left Chatham'.[11]

The French navy's 20-knot 40.75m (133ft 8in) torpedo boat *Balny*, built by Normand in 1886 and seen here off Toulon.

The small wars which raged towards the end of the nineteenth century fostered the production of torpedo boats. Here a batch for Argentina is being fitted out at Yarrow's Poplar yard in 1890; in 1875 Captain Hunter Davidson (who had starred in the Confederates' torpedo boat offensive and later bought Yarrow's first spar torpedo launch in 1872) ordered some Yarrow 16.8m (55ft) torpedo boats for Argentina and started the world-wide convulsion of torpedo boat mania.
W. P. Trotter

Still the addiction to speed was causing problems, hypothetical and real. The torpedo boat was a nocturnal weapon, most successful when stalking its enemy in its lair. For a boat to carry out its attack at 20 knots, when her machinery could be heard and the glow from her funnels seen, was ludicrous. Her handiness and acceleration were assets in escape although only conceived as features necessary for taking the war to the enemy. A dominant reason for the boats being fast was that in daytime attacks they would be able to penetrate the gun-fire of a ship and loose off their torpedoes when inside the arc of fire; with the ship torpedoed the threat from guns would cease. It is barely credible that a daylight attack would have been made but, to support this theory, the propagandists pointed out that in an attack by a dozen boats, at least one would evade the guns and fire its torpedo. Moreover, all twelve boats together cost one-tenth the price of an ironclad. In view of the hash that torpedo boats frequently made of exercises in company it is unlikely that such an attack by ten or twelve boats could have succeeded. Even though 20 knots was a high speed, at its torpedo-firing range of 274-365m (300-400yd) the torpedo boat presented a target easily tracked by quick-firing guns. It was not necessary to sink torpedo boats as they were fairly readily disabled by gun-fire. Realising this, Yarrow constructed an armoured boat, *Kotaka,* 50·6m (166ft), for the Japanese navy in 1885, the largest torpedo boat then built, which led the daring attack on Port-Arthur in the Sino-Japanese War of 1894-5.

Despite the seagoing abilities of Yarrow's boats, the torpedo boats were being built to larger and larger specifications, primarily because of the reputed unseaworthiness of the French boats, which had at least undergone sea trials. In the naval manoeuvres off Toulon in 1886 and 1887 they proved hopelessly unsuited to their dummy rôle of breaking a blockade and boats and crews were incapacitated in the open sea. Two boats capsized and sank in the Mediterranean, and though not ones built by Normand's yard, the *jeune école* of torpedoists found itself in heavy weather, despite the adulation in which torpedo boats were held in France. In 1889 two first-class boats foundered, one off Toulon and the other on passage from Havre to Cherbourg. Normand, like Thornycroft, was conscious of the limitations of his boats, but was unable to prevent the public and naval authorities from charging the boats with duties beyond their capabilities. Normand said that 'should their seaworthiness be proved' then no fleet should sail without an accompanying flotilla of torpedo boats equal in strength to those of the enemy.[12] British torpedo boats were just as susceptible to sea conditions. During evaluation trials in 1887 twenty-four boats were sent on an 80-mile race from Portland around the Ore Stone. A Yarrow boat was first back to Portland, averaging 16 knots, but boiler or engine trouble caused several boats to break down, and in the following year only eight of the twenty-four boats arrived at Portland without any damage. In the same year, in trials evaluating torpedoes, eight of the eighteen torpedo boats broke down when steaming at full power. On fleet manoeuvres in the Channel, the boats, far from leading the ironclads a sprightly chase, reduced the fleet to 10 knots, the most the torpedo boats could manage in a seaway. Their crews spent all their energies on raw seamanship and could no more have aimed and fired a torpedo than steamed at full speed in heavy weather.

The service performance of torpedo boats indicates that the Admiralty was correct in its hesitancy about ordering large quantities of the boats, though its reasons were as much political as naval. Commentators, 'who went no further than to count noses, steadily upbraided our government for its slackness, and shook the increasing numbers of first-class and sea-going torpedo-boats in its face, as if it were obvious that the only answer for the dominant

naval power was to build a greater number of the same class of torpedo boats which the sub-dominant powers were producing'.[13] However, as the following figures show, the Royal Navy did keep pace with Russia and France in total numbers of torpedo vessels. These figures are approximate and include the high percentage of sea-going boats which filled navies' torpedo fleets from 1885 onwards.

	1877	1886	1890	1895	1900
Great Britain	1	130	147	123	95
France	8	107	131	187	192
Russia	26	115	131	116	115
Germany	1	67	107	81	20
Italy	1	70	130	74	71
Japan	—	—	26	55	28
USA	—	1	5	5	12

It was to cope with the threat of sea-going torpedo boats built in Europe which led the Admiralty to look for counter-measures, resulting in seventy destroyers in service by 1896 and ninety-two by 1897.

The answer to unseaworthiness was seen in France as a matter of building bigger boats, and this, coupled with a scaremongering French book by Gabriel Charmes (translated in 1886) and the appointment in the same year of Admiral Aube, a torpedo vessel devotee, as Minister of Marine, led the Admiralty into searching for a vessel that would contain the French boats in their ports, preventing them from crossing the Channel and marauding British ships. There had been desultory attempts to find a suitable torpedo cruiser, capable of bottling the enemy in her ports and sufficiently nimble to evade torpedo boat attacks. HMS *Scout*, 1605·3 tonnes (1580 tons), 67·1m × 10·4m (220ft × 34ft), was built in 1885 with 2834·0kW (3800ihp) engines but her speed of 17 knots was inadequate and, because of her size, she would have been an object of torpedo boat attack, not evasion, while on blockade duty. At the same time J. S. White built the 1007·1kW (1350ihp) *Swift* (later *TB 81),* 127 tonnes (125 tons), 46·6m × 5·3m (153ft × 17ft 6in), a 22-knot turnabout with a crew of 25. It was thought that this might be the type that was needed to accompany a blockading squadron to enforce the destruction of enemy torpedo boats, and its stem was strengthened for ramming and, in addition to three torpedo tubes, six 3-pounder quick-firing guns were mounted. *Swift* was undoubtedly a sea-going boat but was not up to standing off Brest or Cherbourg for weeks at a time. Nevertheless she provided some valuable information for the Admiralty and the Director of Naval Construction, Barnaby, designed a larger version, *Rattlesnake,* 533·4 tonnes (525 tons), built by Laird in 1886. She was a torpedo gunboat, carrying one 102cm gun and six 3-pounders with two torpedo tubes.

Measuring 61·0m × 7·0m (200ft × 23ft) she could only steam at 19 knots, with engines of 2387·0kW (3200ihp). Unfortunately someone in *The Times,* in a fit of chauvinistic inaccuracy had called *Swift* a torpedo catcher (26 November 1885), derived from torpedo boat catcher.

This tag bedevilled all torpedo gunboats from *Rattlesnake* onwards, because it was assumed that their function was to chase, after torpedo boats and give them a hiding. They were designed to blockade, not chase, and when *Rattlesnake* flunked the latter task in trials at Portland in 1887 she, and successive vessels of similar type, were ridiculed for years. These boats were sea-going and had a range of 2800 miles at 10 knots but their inferior speed led the Director of Naval Construction to try for a faster design and in 1888-9 nine *Sharpshooter*-class torpedo gunboats, designed by Sir William White and built by royal dockyards, were launched. They were larger than the *Rattlesnakes,* 746·8 tonnes (735 tons) with 2760·2kW (3700hp), with fewer guns but more torpedoes. Their performance at sea was disappointing and, in the manoeuvres of 1890, one of the four gunboats taking part, *Speedwell,* was almost taken by a torpedo boat and *Rattlesnake* was actually captured by a torpedo boat. To counter this, eleven *Alarm*-class torpedo gunboats were laid down in 1890, with the same machinery but 76·2 tonnes (75 tons) greater displacement in an effort to redress the weak hulls of the *Sharpshooters*. With the exception of *Speedy,* built by Thornycroft with a water-tube boiler, they were no faster than the *Sharpshooters.* In 1892 a fourth attempt was made to find the right formula with five *Dryad*-class ships, 1087·1 tonnes (1070 tons), but the best speed of the fastest, *Hussar*, was 19·7 knots. By this time most navies' torpedo boats were capable of 25 knots. Thus between 1886 and 1892 twenty-nine gunboats were built and pronounced failures without ever being, as Admiral Philip Colomb said, 'tried in their rôle of inshore watchers, but they have been tried and have been understood to fail in the rôle for which they could scarcely have been designed, namely, torpedo boat chasers in the open sea'.[14] In the last ten years of the century, Italy, France and Germany increased the size and numbers of sea-going torpedo boats, which were vessels of sufficient threat to necessitate the counter-measures put in hand by the Royal Navy which resulted in the birth of the destroyer.

At this time there was another outbreak of torpedo warfare during a civil war in Chile in 1891. One faction possessed two 762-tonne (750-ton) torpedo

A Thornycroft first-class torpedo boat of 1880.

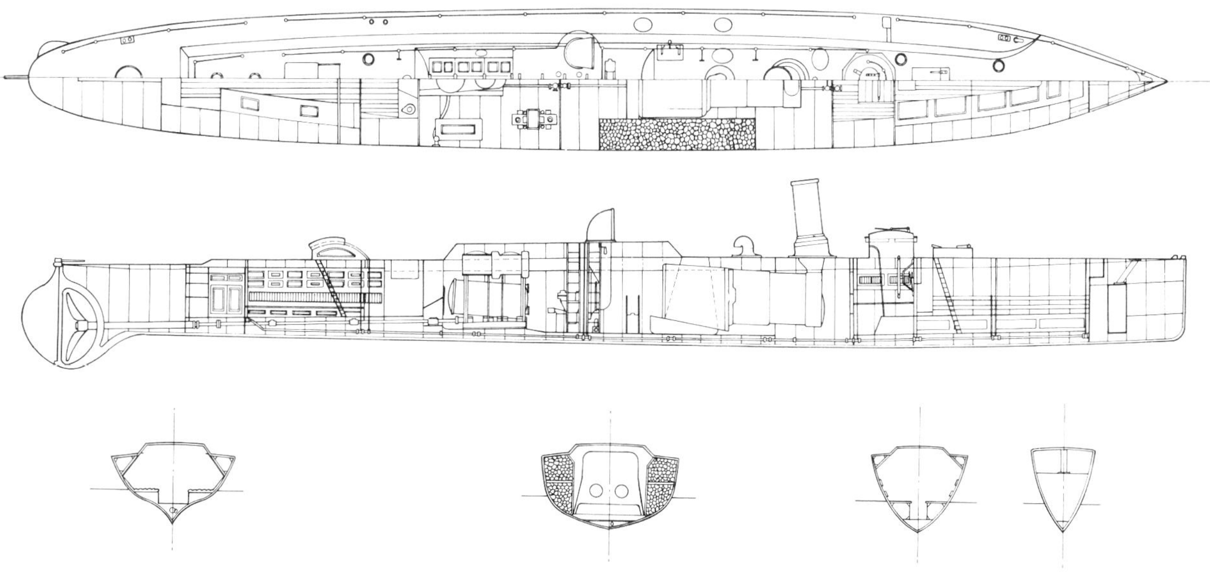

gunboats, *Almirante Lynch* and *Almirante Condell.* On 23 April they closed the ironclad *Blanco Encalada*, 3556·0 tonnes (3500 tons), and *Almirante Condell* fired three Whiteheads which all ran wide of their target. *Almirante Lynch's* first torpedo also missed, at 137m (150yd), but closing further, one of her next two struck the ironclad which sank in six and a half minutes. Then in 1894 a Brazilian torpedo gunboat, the British-built *Gustavo Sampaio,* and three torpedo boats attacked the rebel ship, *Aquidaban.* The torpedo boats did not hit anything but a torpedo from the gunboat hit and disabled the rebel.

In the Far East torpedo boats maintained credibility during the Sino-Japanese War of 1894-5. During the Battle of the Yalu River the Chinese used four torpedo boats, but their torpedo depth-setting mechanisms were badly adjusted for none of the torpedoes fired hit its target, though apparently within range, and the four boats were driven from an assault on Togo's *Naniwa* by gun-fire. The Chinese fleet had received a battering and sheltered for the winter in Wei-hai-Wei. On 13 February, Admiral Ito sent three flotillas of torpedo boats into the harbour. One boat blew up the boom, four others staged a diversion, and three penetrated the harbour in rough and cold conditions and one of their Schwarzkopf torpedoes struck the Chinese battleship *Ting Yuen*, 7620·0 tonnes (7500 tons), which was forced to beach herself. All but one of the Japanese boats escaped after the raid and all were damaged by gun-fire. Early the next morning the Japanese boats made another attack on the harbour and sank a cruiser, an old wooden corvette and a large steam launch, all five boats returning to their fleet unharmed. Though relatively small actions, these skirmishes attracted considerable attention in naval circles. The Japanese attacks provided a useful illustration of the validity of torpedoes in modern warfare and the demand for a counter was increased.

While the experiments with the under-powered and over-sized torpedo were going on, few torpedo boats were built for the Royal Navy and in 1892 Yarrow alerted Fisher to threats from the Continent which had never disappeared: the German torpedo boat builder, Schichau, of Elbing, had a 27·5-knot boat, and Normand was designing a 30-knot vessel. Yarrow suggested the outlines of a suitable antidote to Fisher, a light displacement hull of 54·9m (180ft) with 2984kW (4000hp). Fisher set up a committee in March 1892 to study the proposal, directing that the new vessels, torpedo boat destroyers, should be heavily armed and have a speed of 27 knots. The committee recommended the proposal and in June 1892 orders for two torpedo boat destroyers were placed with Yarrow, followed by orders for two each from Thornycroft and Laird of Birkenhead. In 1893 Yarrow launched the first destroyer, *Havock,* 243·8 tonnes (240 tons), 54·9m × 5·6m × 3·4m (180ft × 18ft 6in × 11ft), mounting one 12-pounder, three 6-pounders and three torpedo tubes. The other ships, Yarrow's *Hornet,* Thornycroft's *Daring* and *Decoy,* Laird's *Ferret* and *Lynx* were all launched by 1894, and though they rarely achieved their speeds (*Ferret* 28·2 knots, *Daring* 28·2 knots, *Hornet* 27·6 knots) in service, the formula had finally been found to combat the torpedo boat. Another thirty-six destroyers were ordered, from fourteen different builders, and in 1895 Thornycroft's *Boxer* achieved a world record speed of 29·3 knots. The destroyers were considered eminently suited to torpedo boat functions and not only usurped their rôle of sinking battleships but were able to do so on inshore waters and the high seas. That they lacked the stealth and shallow draught to launch a torpedo in a harbour was immaterial. The destroyer could overhaul a torpedo boat in any sea and fire at it while out of the boat's range. Their speed was such that in 1895 the crews were issued with goggles to protect their sight from the wind and spray. No navy would breed flies when it owned the swat and in 1895 Yarrow built *Sokol* for the Russian Navy, a 243·8-tonne (240-ton) destroyer of 30 knots, while Normand built *Forban* in the same year, 128·0 tonnes (126 tons), 30·3 knots. Between 1896 and 1901, sixty-seven 30-knot destroyers were laid down and completed; the torpedo boat furore was mute by comparison.

In 1898 Japan ordered destroyers from Yarrow and proved, in the Russo-Japanese War of 1904-5, that the new warships were not going to be as transitory as torpedo boats. The destroyers were able to keep at sea with the battle fleet and played a crucial rôle in the Japanese victory. However the day of the steam reciprocating engine was now drawing to a close. In 1878 two Yarrow boats had flaunted their speed at Spithead and were the darlings of the review. During the 1897 review such torpedo boats that were left in service were tucked away at the back of the fleet.

It had again become almost axiomatic that speed could only be achieved by large high-powered ships. In 1895 the fastest vessel in the world was *Forban,* which used 2965·4kW (3975ihp) to reach her top speed; the 29·6m (97ft) torpedo boat type vessel *Turbinia,* built in 1894 at Wallsend-on-Tyne, was expected by Charles Parsons to be as fast with three steam turbines totalling just 1492·0kW (2000hp).[15] However her best speed was only 19·7 knots because the three propellers were cavitating, that is turning at such a speed that they were creating a froth of air and water and reducing their driving efficiency and power. Propeller cavitation had first been recognised by Jacques-Augustin Normand in the early 1890s when he investigated the problem.[16] He built a hoist to lift the stern of one of his torpedo boats out of a dock with its engines running; he published his findings in 1893. Parsons overcame the problem by distributing the load among three propellers on each shaft and *Turbinia* astonished the Spithead review of 1897 with a speed of 35 knots. In 1899 a steam turbine was fitted in the destroyer *Viper,* which averaged 36·5 knots on trials and topped 37 knots.

The two boatyards which had fostered torpedo boats both left the Thames at the beginning of the twentieth century to concentrate on building destroyers. Yarrow moved to the Clyde in 1907 and Thornycroft, whose vessels were becoming too large to pass beneath the Thames bridges, moved the shipbuilding part of the company to Woolston, off Southampton Water, where it was operational by 1906. HMS *Lightning, TB 1,* was discarded in 1900 and scrapped in 1910.

3 BOATS FOR PEACE

There was little favour for the torpedo boat in the revitalised navy of Fisher. It had been too much discredited and the petrol-engined submarine was the new darling, even in its early rôle as a coastal defence weapon. An alliance with France in 1904 and an arms race with Germany reversed Britain's traditional loyalties and ended the need for small strikers to cross the Channel. The North Sea looked like being the new battle area and ships of the *Dreadnought* and *Invincible* classes were built in readiness to contest Germany's louring navy. The Naval Defence Act 1889 and the withdrawal of the claws of gunboat aggression led to the reconstruction of a capital ship Royal Navy and eclipsed all necessity for torpedo boats. However, the design expertise was still available and the internal combustion engine was an ideal prime mover.

An internal combustion engine had been patented in 1823 by Samuel Brown of London and the following year he built a working road vehicle, but the running costs of the 2-cylinder 2·98kW (4bhp) engine were more than those of a comparable steam engine. In 1827 Brown ran a successful trial on the Thames in a boat fitted with his coal-gas engine, but this and successive attempts by other inventors had no commercial support. Improved gas-engines, notably Etienne Lenoir's engine of 1860, renewed interest but the gas engine was a cumbersome and low-powered machine. The discovery of petroleum in America in 1859, and of its properties as a combustible spirit more efficient than coal gas, led to its use as a fuel for internal combustion engines. The first patent for a petrol engine was granted to George Brayton of Philadelphia in 1874 and J. Spiel took out a German patent in 1883, his design having the benefit of Dr Nikolaus Otto's invention of 1876: the four-stroke gas engine. But these early efforts were little better than the contemporary gas engines and it was the achievement of Gottlieb Daimler to build a lightweight high-speed engine in 1883 that ran on petrol. (Daimler at that time was technical director of the factory at Deutz, near Cologne, which built the Otto engine.) It is interesting to note that he did not consider his engine for land transport and carried out his practical experiments in boats. In 1886 he installed a 0·8kW (1bhp) single-cylinder model in a 5·8m (19ft) boat and ferried eleven people along the river Neckar at Frankfurt at 5½ knots; also in 1886 the motor launch *Marie*, with a 1·0kW (1·3bhp) Daimler engine was presented to Bismarck for use on his lake. Daimler motor car production was not established until 1890. Daimler and Benz engines were fitted to other German boats and in 1890 the boatbuilder Otto Lürssen of Vegesack, Bremen, founded a long-lasting partnership when he installed a Daimler engine in one of his first launches.

In1885 a Londoner, J. Hulme, is supposed to have propelled a boat with an 'internally fired hot-air engine'[1] but there is little evidence to support this pioneering claim. In 1888 the Priestman company of Hull fitted an 8·5m (28ft) boat with a paraffin engine and, by 1890, Priestman engines were propelling barges on the Manchester Ship Canal. J. D. Roots built a petrol engine which he used to run a boat between Richmond and Wandsworth on the Thames in 1891 and 1892; Frederick Lanchester built two successful motor boats in 1895 and 1897 and, also in 1897, the Daimler company of England sold 4·5kW (6bhp) 2-cylinder petrol engines for launches. Hornsby-Akroyd, Priestman, Vosper, Tangye and other engines were installed in small craft but the early engines were not easily managed and their use was largely commercial, in river barges, small ferries and workboats. In America and France the public were less diffident about the internal combustion engine and in America such petrol two-strokes as the Monarch and Wolverine became fashionable and popular in launches and boats. In England the paraffin engine had made some of the early running and it was not until the turn of the century, when Napier and Daimler petrol engines replaced the less than reliable paraffin engine, that marine motoring became practicable, although most boats still required a mechanic. In France the first marine diesel, of 14·9kW (20bhp), had been installed in a canal boat in 1902-3, following the building of the first successful diesel in 1897. One of the engineers responsible was Adrien Bochet, chief engineer of Sautter-Harlé of Paris, the company which built the first diesels for French submarines. French engineers realised that the four-stroke petrol engine had more to offer than the two-stroke and Darracq, Mors, Gobron-Brille, and De Dietrich engines were laying the foundations of a fine tradition.

In France enthusiasm for the petrol engine was manifestly expressed in the large number of motor cars which began to appear on the roads. There were few legal restrictions in France to compare with Britain's locomotive acts (of red flag notoriety) of 1865 and 1878, or the 1896 Locomotives on Highways Act which raised the speed limit to 14mph. The French roads lent themselves to touring in motor cars and until 1903 there was no restriction on racing on public roads. This has never been permitted on the English mainland, and competitive motorists were obliged to get their kicks from racing in Ireland, the Isle of Man and on the Continent. There was considerable resentment towards the motor car, particularly in rural areas, an attitude which the motorists – for the most part young bloods with reckless dispositions akin to Mr Toad's – did little to alleviate. Then, a large number of accidents during the first day of a road race from Paris to Madrid in 1903 caused the French government to halt both the race and all future events on public roads. In the same year in England a powerful lobby (there were about 8500 motor cars in Britain, most of them made in France) was trying to raise the speed limit to 25mph and resist motor car registration. The Motor Car Act passed on 1 January 1904 introduced registration and fixed the speed limit at 20mph. As many motor cars could cruise at twice that speed the effect of this legislation was to accelerate the growing enthusiasm of many men and women for racing motor boats, in which sphere of activity there were no speed limits; indeed fast and reckless driving afloat was laudable by comparison with motoring.

Yachting had been extremely popular in Britain in the 1880s and 1890s. As a sport it had royal patronage, many thousands of participants, and three or four magazines devoted to it; its social aspects and major races were widely reported and accounts of voyages were in healthy demand in the lending libraries. The steam yacht, as built by Thornycroft, Camper and Nicholson, Fay's and Summers and Payne, could be seen in increasing numbers in British waters; in 1873 there were 140 and by 1890 there were over 500. It was not uncommon for some of these vessels to compete together, though more often it was the ships' steam launches which raced privately against each other. Towards the end of the century races began to be held for steam launches at yachting regattas. As early as 1889 the Cercle de la Voile de Paris had been organising events for motor boats. Racing steam yachts had also been popular for a short while in New York, but it was the gradual attainment of moderately reliable petrol engines, and manufacturers and entrepreneurs anxious to prove them afloat, which made possible Sir Alfred Harmsworth's donation of the Harmsworth Trophy in 1903. The Trophy, which soon became known as the British International Trophy, was awarded to the national motor boat team winning a series of races against other countries. It was a prestigious event and the BI races drew together diverse threads in motor boat racing and created a goal for many of the talented designers interested in the sport. The first BI Trophy contest was held at Queenstown, Ireland, in June 1903.

The race was administered by the Motor Yacht Club which had been founded in December 1902 as a branch of the Royal Automobile Club. In the autumn of 1902 Linton Hope, A. F. Evans, and the editor of the *Yachtsman,* E. H. Hamilton, formed the Marine Motor Association, primarily to establish a rating rule for racing; its proposals were adopted by a sister body in America, the American Power Boat Association. The *Yachtsman* magazine, now defunct, was mainly concerned with sailing boats but it recognised the interest in motor boats and reported their progress freely; *The Motor Boat,* 'encouraging and recording progress in the construction and use of power craft', was founded as a weekly in 1904 and in the same year the Royal National Lifeboat Institution sanctioned the installation of a petrol engine in a lifeboat; in the 1908 Olympics in London there were even events for racing boats. Despite the attention surrounding the first BI race only three boats competed and they were all British. The race was won by *Napier I*, a 12·2m (40ft) steel boat designed by the peerless motor boat naval architect Linton Hope. *Scolopendra* and *Durendal* were smaller and less powerful boats and *Napier I* had an easy win. Both Thornycroft and Yarrow were very interested in fast motor boats, and strong advocates of the petrol engine. *Scolopendra* was a 9·1m (30ft) launch, designed on the lines of a torpedo boat by Thornycroft, although the boat was built by another Chiswick yard, Maynard's Motor Launch Company. Her engine was a marine version of the 4-cylinder 14·9kW (20bhp) model Thornycroft built for use in their motor cars and her best speed was just over 15 knots – a remarkable performance for such a low-powered boat. The appearance of the 18·8-knot *Napier I* was in stern contrast to the casual frippery of many of her contemporaries and her fine entry, whaleback foredeck and tumblehome aft set a pattern for racing boats for years to come. She was built by Yarrow for Seaton Francis Edge, better known as a racing driver (he won the 1902 Gordon Bennett Cup Race), who was a partner in the Napier motor car and engine workshop at Lambeth in south London. He took every opportunity to exhibit the prowess of Napier engines, ashore and afloat. *Durendal* was built by S. E. Saunders and had a 37·3kW (50bhp) 8-cylinder engine. *Napier II* was built in steel by Yarrow in 1904 and fitted with two 33·6kW (45bhp) Napier engines, which gave her a speed of 25 knots. Defending the BI Trophy on a seven-and-a-half mile course on the Solent in 1904 she beat the American boat *Challenger,* designed by Clinton Crane, but in a second heat her bow plates opened up and the boat was withdrawn. Edge competed in the final heat in the 10·7m (35ft) 20-knot *Napier Minor* and won, but lost the trophy to the 9·1m (30ft) 23-knot French boat *Trefle-à-Quatre* on a technicality. Though Yarrow was well used to working in steel, the 12·2m (40ft) *Napier II,* with a beam of 1·5m (5ft) and draught of 0·8m (2ft 6in), was too small to warrant the use of metal. She was about three times heavier than a comparably sized timber boat, and the necessity of keeping her scantlings as light as possible resulted in insufficient strength to hold the hull together under the tremendous pounding inherent in racing.

Until the requirements of racing created machinery of low power/weight ratios, powered boats were of conventional displacement form, full bodied with substantial keels and canoe or counter sterns. When such hulls are charged with more power than they can convert into motion, their wave-length increases, the stern starts to sink in the trough of the bow wave and the hull becomes unstable and incapable of going faster. The slender torpedo-boat hulls had shown one way of coping with the problem but the racing boats were pushing forward power/weight ratios to unknown limits. A solution soon reached was to flatten out and widen the afterbody of the hull to provide a large area on which the boat could squat while under way at speed. The external keel was eliminated and engines mounted as low as possible to keep propeller shafts near horizontal. Bow sections were fine and midship areas became artfully designed and superbly constructed amalgams of two wedges; the leading one vertical and the after one horizontal. Like many sudden developments, this teetered towards self-parody, with forebodies honed like hatchets and the freeboard and draught aft measuring just a few inches. This form made high speeds possible, though not at sea. It was this limiting aspect of the design which led N. G. Herreshoff to abandon work on his light steam launches of 1876-8, which were in some demand by the US Navy, Ordnance Department and Fish Commission, and useful vessels in open water. But remarkably, an uncompleted model for a torpedo boat from that era constituted the design for a Herreshoff boat of 1904, *XPDNC*, a 13·1m (43ft) racer with the same twist in hull shape which was evolving in Europe. She had a 56kW (75bhp) Mercedes engine and a speed of 28 knots. Though she won the National Trophy in 1905 she was not raced in Europe, and in America her career was eclipsed by the boats of Clinton Crane. Crane had designed the American entry in the 1904 BI Trophy, *Challenger*, and it was another Crane boat, *Vingt-et-un,* which beat Herreshoff's 15·9m (52ft) steam launch *Swiftsure* in a race of 1904 which made it clear to the reluctant Herreshoff that the steam boat had been displaced. Crane's boat also ended the British and French domination of the BI Trophy, when the Motor Boat Club of America offered a challenge for the 1907 race with a boat called *Dixie*, which won at a speed of 27·5 knots.

In *Napier I* Hope devised an unusual means of construction, an idea implemented in many high-speed boats and which culminated in the coastal motor boats of the First World War. He lengthened the normally stubby engine-bearers to form the main longitudinal strength, running from stem to stern. With the strength of the hull embodied in powerful girders savings in weight could be made elsewhere, such as in dispensing with the keel. *Napier II* was similarly constructed, as were two other famous Hope designs. *Legru-Hotchkiss* was an 11·9m (39ft) racer for M. Legru of Paris. It was built in his own workshops in 1904 by boatbuilders from S. E. Saunders' firm. With two 63·4kW (85bhp) engines she achieved 29·68 knots over a five-mile trial on the Seine in 1905, taking the speed record from the wholly French boat *Dubonnet*, a formidable 15·2m (50ft), 223·8kW (300bhp), Tellier-Delahaye racer which had dominated the competitions at Cannes and Monaco with a best speed of

28·1 knots. J. E. Hutton's *Hutton II* had a 6-cylinder Hutton engine (the degree of commercial vanity then being almost as widespread as today) which developed 104·4kW (140bhp) at 1150rpm. She was very strongly built in timber, which Hope realised was far better for fast boats than steel, with three 5·1cm (2in) pine girders, 40·6cm (16in) deep midships, running the length of the hull. She could reach 25 knots but persistent engine trouble deprived her of success. She was bulbous in appearance, in contrast to lean contemporaries, and was at her best in a seaway.

By 1905 Hope's principle had advanced to the stage where cross-braced longitudinal girders were an integral structure, contributing little to the strength of the outer hull, serving only as a frame on which a skin was attached which needed to be no stronger than local stresses required. This technique permitted the construction of the stepped boat, the skimmers which were to dominate fast boat design for ten years.

It is often supposed that the hydroplane, in single-step and then in multi-step form, was iconoclastic. Temporarily putting aside the early influence of Ramus, Pictet and Thornycroft, it is apparent that without the first racing boats such as *Dubonnet, Panhard-Levassor, Mercedes-Charley, Challenger, Baby II, Hutton II* and the Napiers, with tight small-radius bilges, the advent of the hard-chine planing boat would have been deferred. Hope's boats in particular were all but planing craft in intent and when *Napier II* was redesigned at the end of 1904 she embodied elements crucial to the planing form. When her bow plates sprung in the 1904 BI Trophy she was strengthened, but inadequately; she then underwent trials with various types of fore-body. After tank testing and when models had been towed from a torpedo boat, in a final spurt of ingenuity Hope and Messrs Yarrow cut the forefoot right away and raised the first 4·6m (15ft) of the hull to an angle of attack of about 4 degrees, with the intention of lifting the fore-body clear of the water when underway to reduce skin resistance. A commentator noted of a sister ship, *Yarrow-Napier,* that 'Although it is difficult, from a mechanical standpoint, to estimate the advantages gained by the adoption of this form, there appears to be little doubt that the "skating" effect has been achieved'.[2] The boat was also fitted with two 3·7m (12ft) long spray deflectors rivetted along each side of the foredeck. With little or no sheer racers were very wet boats, but *Napier II* had pronounced sheer in her hull, and it is tempting to consider that the spray guards were forty years in advance of the work eventually done in generating lift with spray rails on the hull. (Thornycroft torpedo boats built for Russia in 1904 also had spray strakes from the bows to midships.) In February 1905 the modified *Napier II* was taken for trials on the Thames, topping 27 knots with the tide and 25·9 knots on mean. Alfred Yarrow explained, simplistically, that the boat was borne in a different way when at high speed than when at low speeds: 'The boat wanted to sink into the water, but had no time to displace the water, and consequently did no more than skim the surface.'[3] In that summer *Napier II* won the BI Trophy at Arcachon in France, at 22·5 knots. It would seem that although the hard-chine hull per se had its origins in the minds and models of Ramus and Thornycroft, practically it was developed from the racing boats of 1903-8. Nevertheless a similar approach had been studied in France, where Maurice Le Las had realised that the simple punt form could be made to plane and, perhaps as early as 1904, he had applied Ramus's idea of a break, or step, in the hull to produce what he called a ricochet boat, though it was not until 1908 that the idea achieved racing success and public acclaim. However an engineer and inventor, M. Bonnemaison, entered his 3·5m × 1·8m (11ft 6in × 6ft) hydro-glisseur, powered by a 6·7kW (9bhp) V2 Deckert, for the race meeting at Rouen in

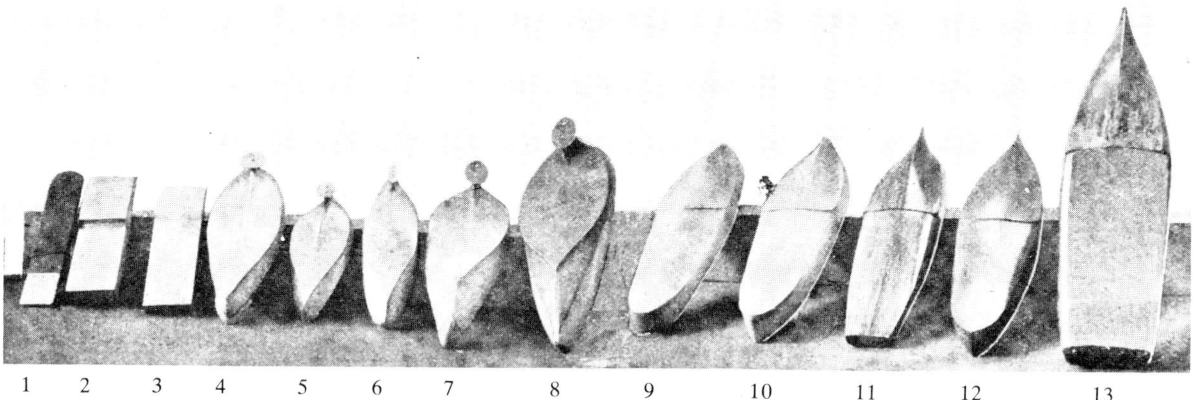

The complete range of models of Thornycroft's skimming hulls, from his first *(1)* of about 1873 to the stepped Miranda type of 1912 *(13)*. Models *2* and *3* are planing hydroplanes of the Ramus period and *4* to *7* are experimental hulls with planing foils at the bow which led to *Miranda III (8)*. The prototype for the stepped hull of *Miranda IV* is *9* while *10* is the actual model from which she was built. Number *11* is a model of *Maple Leaf III*, *12* an experimental model while *13* is Dan Hanbury's 1912 BI Trophy challenger, designed by Thornycroft but built by Luke & Co, Hamble. *Motor Boat and Yachting*

1906. The first hydroplane victory in a motor boat race was on 14 October 1906, when de Lambert's entry won the Auto Cup in Paris, a 100km (62·1-mile) event. *Ricochet II* and *Ricochet-Nautilus* were also competing but spray stopped their engines.

A stepped boat is one in which the planing area of the hull is focussed on the vicinity of the step. Once the boat is travelling at planing speed the hull rides on the step, considerably reducing the total wetted surface and making possible quite high speeds with moderate power plants. Provided the engine can overcome the low-speed inefficiency of the hull, and the drag of its step, and propel it beyond its displacement speed into planing speeds, the small hydroplane can attain impressive speeds in sheltered waters. The racing hydroplane became popular in Britain, being easy to build and fun to race, and the Motor Yacht Club created a special class for hydroplane events in 1908.

Thornycroft meanwhile was evolving a type of hull which could be used at high speed at sea. With lightweight engines available, some produced by his own company, Sir John Thornycroft was able to put into practice the idea for planing craft he had conceived thirty years earlier. In this he was helped by the enthusiasm of his son, Tom Thornycroft (whose elder brother J. E. Thornycroft took over the running of the company in 1901). Tom Thornycroft used to race the firm's motor cars and boats, notably *Scolopendra* and her successor *Gyrinus*. This boat was built by Maynard, but in 1908 Thornycroft attached a planing surface to the hull in the form of horizontal fins above the chines, to gain lift. This generated a small but significant increase in speed to 21·7 knots and *Gyrinus* won two of the three 64km (40-mile) events in the Olympics. In 1909 Thornycroft built *Miranda III*, also a flat-bottomed boat, of 6·7m × 2·1m (22ft × 7ft) with a 44·8kW (60bhp) engine, fitted with a steel plate, in the shape of an inverted T, beneath the keel at her bow. At speed the boat rode on her flattened afterbody and the bow foil. Though commonly regarded as one of the first successful hydrofoil boats, inventors in France, Italy, America and Britain had been experimenting with fast craft which travelled on appended foils for over thirty years. Many designs were never implemented but the boat of Comte de Lambert and Horatio Phillips reached 8·6 knots in 1897. It was a steam powered catamaran which rode upon four in-line angled foils, and by 1907 de Lambert had achieved 29·5 knots. In the same year the Italians Crocco and Ricaldoni claimed 43·5 knots for their 7·9m (26ft) hydrofoil. Another Italian, Forlanini, had reached 38 knots in his 1906 hydrofoil and an American, Peter Cooper Hewitt, reached 26 knots in 1907. (For a full account of these turbulent experiments see H. F. King's *Aeromarine Origins*.) The hydrofoil attracted considerable inventive skills, but Thornycroft's contribution in *Miranda III* was to embody the data in a practical boat. Until then hydrofoil craft had been rather spidery (and still are to some minds) and most were driven by airscrews.

It is interesting that while hydrofoils had a certain recondite following, there was equal interest in air-cushion craft. J. I. Thornycroft had patented a design for a hull riding on a body of air encased between hull and water in 1877 and revived the idea in 1908. The work of the American Clement Ader, as early as 1896 and materialising in 1904, was concerned with air pumped into cavities beneath the hull.

The unwieldiness of the hydrofoil averted commercial interests and this led de Lambert to embody planing foils directly on the hull, in the manner of hydroplanes. At the conclusion of his hydrofoil experiments he designed another steam-driven catamaran in 1907 in which there were 'five box-like transverse floats, having punt-like "bows" and mounted on two longitudinal members'.[4] The prototype had an airscrew but successive versions had petrol engines and water propellers. Though not strictly a hydroplane, because the hull rode on a single plane instead of one or more points, de Lambert's *glisseur* undoubtedly commended itself to the American, W. H. Fauber, then resident in Paris. Fauber, a Chicago industrialist, established a boatyard at Nanterre to build hydroplanes to his own 1908 patent: multi-step V-hulls of fairly constant deadrise. He produced 4·9m, 7·9m and 14·9m (16ft, 26ft and 49ft) versions and claimed speeds of between 35 and 40 knots with a 373·0kW (500bhp) engine in the largest model. Fauber's boats were rather more sea-kindly than the stepped hydroplanes of Le Las, Bonnemaison, and de Lambert, and they frequently achieved racing success with considerably less powerful engines than their displacement hull competitors.

S. E. Saunders started his boatyard at Goring-on-Thames in 1870, and in 1884 began to consider the possibilities of laminated hulls, whereby thin strips of mahogany and cedar are laid and fastened over each other, building up strength with lightness and allowing curves denied to conventionally planked boats, even of double diagonal construction. There were usually between three and eight veneers, each mechanically sown with copper wire to the veneer beneath, permiting a saving in weight of up to 40 per cent. The method achieved fame when Saunders built a 23·8-knot umpire's launch for the Henley regatta in 1896, *Consuta,* which gave its name to the system. He was also one of the various enterprising boatbuilders scattered around the country who experimented with petrol engines, having installed a small Daimler engine in one of his launches in 1893. In 1900 the firm moved to Cowes, where in 1903 he built *Durendal,* a challenger for the inaugural Harmsworth Trophy, and the Goring yard was gradually run down. Saunders was also exploring new ground in construction and design: his 12·2m (40ft) *Napier* of 1905 had watertight bulkheads and carried the engine crankcase in a long cigar-shaped bulb along the line of the keel to lower the centre of gravity and provide a flat, almost skimming bottom; Augustin Normand prepared a similar design in the same year.

Saunders built his first hydroplane in 1908, a 7·9m (26ft) single-step boat which, if not a great success, at least demonstrated to Saunders the possibilities inherent in the hydroplane. It was Saunders who built the last fast displacement boat of this era, the splendid *Ursula,* and the first successful racing hydroplane, *Pioneer. Ursula* was the Duke of Westminster's successor to his *Wolseley-Siddeley* of 1908, the first motor boat to reach 30 knots; and in *Ursula,* powered by two 12-cylinder 279·8kW (375bhp) Wolseley engines, Saunders built him a boat capable of 35 knots. *Pioneer* was built a year later in 1910, and sounded the end of displacement boat racing, so obviously superior was the racing performance of the hydroplane. Saunders reached an agreement with Fauber to build multi-step hydroplanes in England under licence, though Saunders made better sea boats of them. He reduced Fauber's seven or eight steps to five, substituted a raked forefoot for his flat forefoot and drop-keel forward, and deepened the slightly concave bottom sections for nearly the whole run of the hull; but he stuck to Fauber's principle of giving each step equal power. A comparison of the two boats is relevant: *Ursula* was 14·9m × 2·0m (49ft × 6ft 6in), displaced 5·1 tonnes (5 tons) and had a speed of 35·3 knots with 559·5kW (750bhp); *Pioneer* was 12·2m × 2·0m (40ft × 6ft 6in), displaced just over 4·1 tonnes (4 tons) and with a single 268·6kW (360bhp) Wolseley engine achieved 40 knots. *Pioneer's* performance was instrumental in popularising hydroplanes and her success led Saunders to concentrate on the multi-stepped hull. It was the speed of *Pioneer* which led the Motor Boat Club of America to defend the BI Trophy with a hydroplane. *Pioneer* was a faster boat than *Dixie II,* an 11·9m (39ft) boat of 31 knots with a 164·1kW (220bhp) V8 engine, but during the 1910 challenge at Huntington Bay, New York, *Pioneer* overplayed her speed, got spray on her ignition system, and lost the race while drying out the electrical circuit. Clinton Crane's single-step hydroplane was *Dixie IV* and she successfully defended the Trophy in 1911 against *Pioneer* in similar circumstances. However, Crane was not happy with the handling characteristics of hydroplanes in rough water and he stopped designing them after *Dixie IV*. *Ursula* was not eligible for the BI Trophy, for which 12·2m (40ft) was the maximum length, and though she dominated the racing in the Mediterranean for a couple of seasons and epitomised the second generation of high-powered boats, Saunders' greatest success, *Maple Leaf IV,* was a multi-step hydroplane developed directly from *Pioneer.*

The only known photograph of *Pioneer,* Saunders' multi-stepped hydroplane of 1910, whose racing success led to the widespread popularity of stepped hulls.
Jon Bannenburg

At about the same time that Saunders was investigating the Fauber design, Linton Hope was also studying it. He produced a design in which the steps followed the normal lines of the displacement racing launch, instead of the hollowed sections of the Fauber patent. This was also the line taken by Thornycroft, who, not attracted by the multi-step principle, plugged on with experiments to produce sea-going planing craft. In this the company was stimulated by the 1906 patent of A. E. Knight (No 17,360) who arrived at his idea for a stepped hull independently of Ramus or Thornycroft. (Knight was Lord Howard de Walden's engineer aboard his 1908 BI Trophy challenger *Daimler II.)* Between 1909 and 1910 Thornycroft abandoned research on hydrofoil craft such as *Miranda III* and designed and built *Miranda IV,* launched in 1910. She was a 7·9m (26ft) single-step hydroplane of 35·5 knots, built on the lines suggested by Hope, with the step hugging the lines of the hull. Her immediate antecedent was a similar boat called *Zigorella,* a 1910 BI Trophy competitor designed by Thornycroft for Dan Hanbury but built by another yard. She had a fair displacement entry until the step in the hull amidships, and a flat afterbody. *Miranda IV* combined this step with the chines of *Gyrinus,* which had in effect been brought inboard to comprise an integral part of the hull and provide considerably more lift than was available to *Zigorella,* with softly curved forward sections and a flat afterbody to form the equipollent plane. *Miranda IV* represented an important stage in fast boat design and a climax to J. I. Thornycroft's personal interest in the subject, for she had the ability to skim over the surface. Thornycroft had tried for this effect with *Lightning* but failed for want of suitable machinery. Like the first *Miranda, Sir Arthur Cotton* and *Lightning, Miranda IV* was a light, easily driven boat. She displaced about 1·0 tonne (1 ton) and her performance with an 89·5kW (120bhp) Thornycroft V8 engine was amazing when compared with the larger and more powerful *Ursula* and *Pioneer.* She was raced with consistent if moderate success and was notable for her ability to keep going and keep dry in a seaway. Two elongated engine-bearers provided the main strength, as in Hope designs, and the step itself was added to the outer skin of the hull, preventing any weakening break in the boat's longitudinal strength. Saunders' multi-stepped boats called for masterly shipwrighting because each step was wrought as part of the hull. In any heavy weather which *Miranda IV* might reasonably be expected to endure, as a lightweight, fast launch – sea states 4 or 5 – she could run at up to 29 knots without too much discomfort. This was largely the contribution of the form of her step, which followed the curved section of the hull from keel to chine. By not allowing the step to compromise the seaworthiness of the hull, Thornycroft created a type of boat which was to achieve success in many areas.

By 1910 motor boats had achieved a considerable degree of reliability, though not without a bashful awareness among their users of the low regard in which they had been held. The Marine Motoring Association was sufficiently anxious about the state of the sport to initiate annual reliability trials in 1904, which were administered by the Motor Yacht Club at Poole from 1905 onwards. In 1906 another annual event started, a rally from London to Cowes. Builders were producing motor cruisers and launches which gained from the expensive lessons learnt by rivals such as Yarrow, Brooke Marine, Saunders, Summers & Payne, Thornycroft, Hart & Harden. Racing boats were often built in the expectation of absorbing experience – and publicity – by putting ideas or equipment through the exacting mill of competition. Commenting on a private venture by Yarrow, Jane's *Fighting Ships* remarked, 'This motor boat is said to possess excellent sea-going qualities, being strongly built and not to be looked upon in any sense as coming under the category of a *racing boat,* which is generally built to win a race and then possibly collapse'.[5] That was a 1906 point of view, and already slightly passé, when by the time of *Miranda IV,* the design of motor boats had progressed from the sound bedrock of knowledge gained by extensive tank-testing and racing. Saunders' application of Fauber's 1908 and 1909 patents resulted in magnificent racing boats, one of which wrested the BI Trophy, yet despite some rise of floor and concave sections, Saunders-Fauber boats were inherently hard-chine flat-bottomed craft with limited seagoing abilities. The pre-eminent Saunders boat was *Maple Leaf IV,* built for Sir E. Mackay Edgar in 1912. His *Maple Leaf III* was a Thornycroft design of 1911 built under licence by Dixon Brothers of Southampton. Driven by T. O. M. Sopwith, *Maple Leaf IV* won the BI Trophy races of 1912 and 1913. The boat was 12·2m × 2·6m (40ft × 8ft 5in) and, powered with two 298·4kW (400bhp) 12-cylinder Austin engines, was the first boat to reach 50 knots. To compensate for the steps in the hull she had a single lattice girder running along the centre line of the hull from the transom to a point 1·5m (5ft) from the bow.

A feature of *Maple Leaf IV, Miranda IV* and a few other progressive designs of this time was their slightly concave bow sections, described as wave-collecting and thought to help the boat ride upon their bow waves. In an attempt to overcome the pounding which flat-bottomed boats encountered at sea, Messrs Cox and King, who built launches and cruising boats at their yard in Wivenhoe, Essex, evolved a stepless chine hull with rather steeper deadrise than was normal, and concave bow sections. This was largely the work of naval architect F. Gordon Pratt, and his designs – of ingenious compromise in the matters of speed, ride and seaworthiness – were revived after the First World War by dozens of fast-boat builders. A notable Cox and King design was *Fascination,* a 20-footer, built in 1912, capable of 24 knots with a 33·6kW (45bhp) engine. In North America the principle of a hull riding on its own bow wave was taken a stage further by Albert Hickman of Canada. His sea sleds were punt-like boats with inverted-V tunnels in the centre of hull, flattening out towards the stern. Often driven with surface propellers, they were designed to cushion the hull on impact with the surface. They were popular, though susceptible to pounding and difficult to construct until reinvigorated in glass fibre as dories.

At first there were no offensive motor boats in the British, French, German, Italian or American navies though there had been sporadic clamouring for armed motor boats even in their days of unreliability. The arguments put forward in favour of their naval use were strikingly similar to those of thirty years before advocating mosquito fleets able to sting the enemy under cover of darkness and escape at speed. Paraphrasing Alfred Yarrow, *The Motor Boat* commented:

'One can see changes at hand in smaller vessels. Take, for example, the torpedo boat. To get 30 knots a boat [61·0m] 200ft long was necessary. This speed will probably very shortly be achieved with a boat, perhaps [18·3m] 60ft long, and she will be light enough to be lifted out of the water and carried on the deck of a battleship. The consequence can easily be seen. Instead of large sea-going torpedo boats cruising independently, battleships will carry two or three small torpedo launches, which can be put afloat when required, or slung inboard if the weather is too heavy for them. These torpedo boats will be worked by three or four men only, instead of a large crew, and they will be the more deadly through being so small that to hit them with a shell would be almost an impossibility.'[6]

By this time torpedo boats, as originally conceived by Thornycroft and Yarrow, had developed into destroyers, and none of the major navies had many torpedo boats of any fighting value. In 1905 Thornycroft had built a 12·2m (40ft) petrol-driven launch, *Dragonfly*. Armed with one torpedo tumble-launched from a side cradle, her speed was 10 knots, and though refitted with a larger 69·4kW (93bhp) engine in 1908, she was not commissioned. Yarrow also hoped to resuscitate the small torpedo boat and, as a private venture, built one which was 18·3m × 2·7m × 0·3m (60ft × 9ft × 1ft), and launched in January 1906. Derived from *Napier II* she was built in steel and powered by five 56·0kW (75bhp) Napier engines of 13 litres each. The engines were cast in gunmetal and the port and starboard pair were arranged in tandem; the fifth unit drove the central propeller shaft and, fitted with reverse gear, was primarily for low speed running and manoeuvring. This boat was known for some time just by her yard number, *1176*, while she was tested and evaluated by her builders. Her trial speed was 26·1 knots but she was running light and displaced less than 8·1 tonnes (8 tons). Yarrow hoped that her performance would recommend itself to the Admiralty, and in August 1906 she was commissioned, but as a picket boat, *Mercury 2,* not a second-class torpedo boat. Service trim reduced her speed to 19 knots. The Royal Navy was considered by some to be languid in adopting the petrol engine for its boats (even though petrol engines of 111·9kW to 410·3kW [150bhp to 550bhp] were used for surface propulsion on the first thirty British submarines produced between 1902 and 1906) but it felt that no fuel with a flash point below 132°C (270°F) was safe. Though the original design of *Mercury 2* made provision for mounting on the stern deck a single torpedo-tube which could be trained on either beam, the Royal Navy was not prepared to engage a petrol-engined boat in warfare. Nevertheless, the advantages of the Napier installation were apparent. Compared with a similar sized steam torpedo boat, *Mercury 2* was two-thirds of the weight, 6 knots faster, and had a radius of action of 250 miles with 1·0 tonne (1 ton) of fuel (stored in tanks on deck), while the steam boat required 2·0 tonnes (2 tons) for a radius of 108 miles. The petrol engine was

Dragonfly, **Thornycroft's steel 12.2m (40ft) motor torpedo boat, ordered in 1904 by Russia but the war with Japan had been lost in August 1905 before the boat could be delivered.**

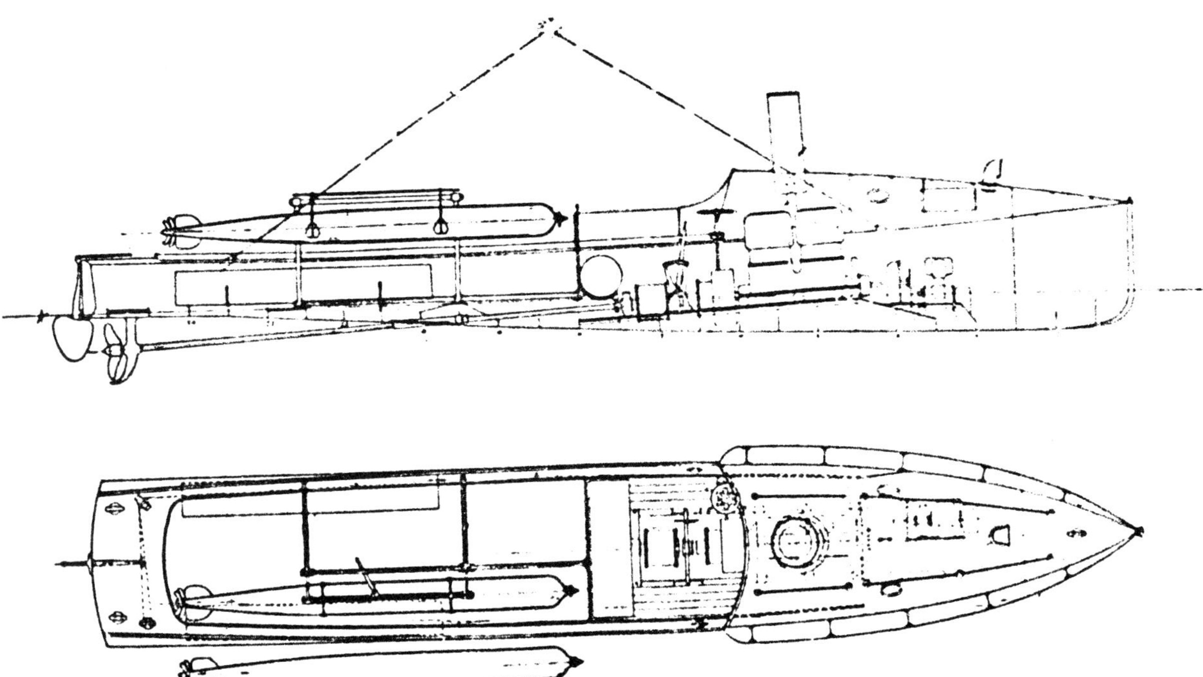

noisier than the steam engine, but did not emit smoke, was ready for sea much sooner, and needed half as many men in the engine room. *Mercury 2* was in constant use as a picket boat and helped dispel such feeling as existed against petrol and petrol engines. In addition to the more acceptable paraffin engines, they were fitted in increasing numbers in barges, ships' boats, tenders and harbour launches. HMS *Dreadnought* carried two 15·2m (50ft) motor launches at her commissioning in 1906. Yarrow even succeeded in selling two of his 18·3m (60ft) motor torpedo boats to the Austro-Hungarian navy. By 1907 labour problems (indigenous on the Thames below Tower Bridge) had forced Yarrow to leave Poplar, and the successor to *Mercury 2* was built at the new yard at Scotstoun on the Clyde. Known as *1225*, she was faster and larger at 27·4m (90ft) long, and powered by four 6-cylinder 149·2kW (200bhp) Napiers but she was not commissioned.

Though *Mercury 2* was the first boat of her type in the Royal Navy, it is possible that the idea, or at least the impetus for her, came from America, where the Standard Boat Company of New Jersey built a steel motor boat designed by Lewis Nixon. *Gregory* was 27·4m × 3·7m × 1·2m (90ft × 12ft × 4ft) and powered by two 223·8kW (300bhp) 6-cylinder Standard petrol engines. She was the first of a dozen boats ordered by the Russian navy to assist in the torpedo war against Japan. She accomplished a voyage across the Atlantic in 1905, supposedly to compete in the racing at Monaco, but instead continued on to Sevastopol. Her range at a full speed of 20 knots was 1400 miles and at 10½ knots, 3300; fuel capacity was 18160 litres (4000 gallons) in seven tanks. Nixon was aboard *Gregory* during the crossing but his remaining boats were pre-constructed in America and shipped to Russia for assembly. A number of smaller petrol vedette boats were also built in France and Belgium for the Russian navy.

No sooner had the petrol engine become a practical accomplishment than a strong lobby urged its widespread adoption, but in Britain the Royal Navy was not prepared to be harried while the petrol engine was unreliable. 'The motor boat for naval use is being industriously puffed, but even in its present

Mercury 2, **built by Yarrow in 1906, was designed as a motor torpedo boat but she was only taken into service in the Royal Navy as a picket. In this 1910 photograph a deckhouse is in the position of the original petrol tanks but two 24-knot boats of this type were sold to Austro-Hungary fitted with torpedo tubes on their stern decks.** *National Maritime Museum, London*

form its utility is yet undemonstrated, save perhaps for harbour service.'[7] The Royal Navy was right to wait. Nevertheless, in Italy and France naval officials were prepared to back motor boats for rather more gruelling duties than harbour service. In 1906 the French navy offered prizes for the best paraffin engines designed for fishery protection vedette boats, and in 1905-6 Count Récopé designed a steel 16·2m (53ft) 8·1-tonne (8-ton) motor torpedo boat of 20 knots, fitted with a torpedo tube in her bows, which was attached to the navy at Cherbourg in 1907. In Italy in 1906 Fiat joined forces with the La Spezia shipyard of Cantieri Navali di Muggiano to sire Fiat Muggiano, which was run by Major Laurenti, the submarine designer and builder. The first boat was launched in April 1906, a 15·2m (50ft) galvanised steel second-class torpedo boat with two 74·6kW (100bhp) Fiat petrol engines. These gave a speed of 15 knots, which was reduced after two Maxims and a 47mm gun had been fitted. A yard model of the boat shows two lightweight torpedo tubes, one mounted on either side of the wheelhouse, and aiming about 10 degrees off the centre-line in the classic position of Second World War motor torpedo boats (MTBs). April 1906 was also the first time a motor boat was used in naval manoeuvres. An 8-knot, 8·5m (28ft) cabin launch powered by a 6kW (8bhp) Union engine was used to represent an offensive torpedo boat destroyer during Royal Australian Artillery trials off Sydney. A German boatyard also built a petrol motor boat in 1907, capable of carrying a 35·6cm (14in) torpedo, and Wilhelm von Siemens was experimenting as early as 1906 with torpedoes controlled remotely by wire. This proved unsuccessful and he transferred his equipment to fast motor boats and was able to control the Lürssen-built *Havel,* which had two 74·6kW (100bhp) petrol engines, at speeds of up to 34 knots. Lürssen built other boats for cable control, and in 1909 the Imperial Navy ordered the 9·1m (30ft) *Racker,* a wire-controlled boat of 28 knots, for trials. In 1911 another German, Wirts, demonstrated unsuccessfully a remote control boat to military observers, and in 1913 Lürssen built *Boncourt,* a 9·5m (31ft) boat of 43 knots, which in 1916 was equipped to fire torpedoes. (Wire-controlled weapons passed out of favour after the First World War, but wire-guided torpedoes and missiles are currently in use with western powers, notably Germany, and also Sweden and America.)

At the outbreak of the First World War none of the major navies had any offensive motor boats in service. Most of the Royal Navy's 109 torpedo boats were over 101·6 tonnes(100 tons) displacement and nothing smaller had been ordered since 1889. America had 24 large torpedo boats and one second- and one third-class steam torpedo boat. France had 17 seagoing torpedo boats; Germany had 45 torpedo boats of between 132·1 and 172·7 tonnes (130 and 170 tons), one of which – built in 1889 – was re-engined with diesels at the beginning of the war; Italy had a handful of 39·6-tonne (39-ton) torpedo boats plus the Fiat vedette; and Russia had 18 torpedo boats of 121·9-203·2 tonnes (120-200 tons) and ten 101·6-tonne (100-ton) motor boats, not known to be armed but possibly the residue of the *Gregory* class ordered in 1905. That some powers had sensed, or heeded, the necessity for small motor boats during war was confirmed by the speed with which Italian and German boatyards built large numbers of high-speed torpedo boats.

4 BOATS FOR WAR

The First World War was remarkable for the intensity of small boat activity in the North Sea and English Channel. No naval war had been waged so close to English shores for a century and the Navy had considerable need of small craft. Moreover it had realised this for many years prior to the outbreak of war and had made considerable provision for training and equipping a domestic fleet of boats. As early as 1906 there had been a call for a coastal reserve by Sir William Haynes Smith. Then, the Admiralty had thought that such an organisation might have its uses but it was not a case for official recognition. In 1907 Admiral Beresford noticed the large numbers of fishing trawlers in Grimsby and as a result of his proposals the Trawler Reserve was founded in 1911 as a division of the Royal Naval Reserve. The first units were in Grimsby and Aberdeen and by 1914 there were eighty trawlers ready for adaptation to minesweeping duties. Other individuals, notably Arnold Foster, seeing the growing numbers of seaworthy motor boats and enthusiastic owners around the coast, made similar suggestions, not uninspired by Erskine Childer's German-baiting novel *The Riddle of the Sands.* Published in 1903, this hugely popular book was instrumental in directing public attention towards Germany and alerting the authorities to the possibilities of small boat warfare. Responding favourably to one suggestion in 1911, an Admiralty memo said, 'These motor boat men have got plenty of money, and we can give them almost anything except money.'[1] However, this support was not always gracious; when some motor boat owners in the Firth of Forth wanted to organise themselves into a local militia, the Admiralty thought that if owners placed 'themselves and their craft readily and unreservedly at the disposal of the local Naval authorities in time of emergency it might very possibly be of assistance'.[2] Representing some 3000 members alone, the British Motor Boat Club suggested in 1912 that owners might form a division of the Royal Naval Volunteer Reserve, and in November 1912 Admiral Inglefield reported that motor boats could fulfil a useful patrol service in estuaries and coastal waters. By February 1914 the Admiralty had approved the formation of a Royal Naval Motor Boat Reserve, though not without some internal wrangling, and a training scheme was put in hand.

But during the first months of war it was the force of armed trawlers and drifters that was able to deal with the German minefields which threatened merchant shipping and the Grand Fleet so severely. By September 1914, 250 trawlers had been armed with two 3-pounders and minesweeping or anti-submarine gear. They patrolled Scapa Flow and key areas of the Channel and North Sea. The trawlers were of incalculable value keeping sea lanes and approaches free from mines. The sturdy boats, and even sturdier crews, frequently had to carry out sweeps in the most appalling conditions. Once the Naval overlords realised their inherent quality and bravery, they were less stringent in exacting naval discipline and achieved even better performance from Harry Tate's navy. With minesweeping in hand, attention was then directed towards the prevention of enemy minelaying. Scores of yachts were on offer to the Navy, dozens more requisitioned, and from 2 September 1914, yachts from all over the country were sent to Portsmouth and Devonport for commissioning, manning and arming with 3-pounders or 6-pounders. These vessels were formed into the Yacht Patrol, manned with a mixture of their yachting crews, fishermen and Merchant Navy men. Officers were usually retired naval officers and Royal Naval Reserves volunteers; the owners were not encouraged to stay aboard and it is probable that few could have stood the sight of their yacht's brightwork being slapped with grey paint or the teak decks being mounted with guns that shook every timber when fired. By 1915 there were 825 vessels in the Yacht Patrol, stationed at Dover, Tyneside, Humber, Loch Ewe, Cromarty and the Shetlands.

Many of the boats offered to the RNMBR after the war started were not suitable for much except the lightest of duties but the more seaworthy boats, with RNVR skippers, were attached to the Yacht Patrol stations and used for picket duties, ferrying, harbour defence and policing, while fitted with little more than small arms: the Navy's watchdogs.

The RNVR crews shaped up into a tireless fighting force, handicapped only by the limitations of their craft; after a year of continuous service many of the smaller boats were of little value. In July 1915 the RNMBR was amalgamated with the Yacht Patrol, and the entire little ship navy numbered almost 1000 vessels: fishing boats, paddle steamers, yachts, motor boats and launches. The Admiralty had been aware of the increasing deficiency of the motor boats and for some time had been anxious to supplement the ailing fleet, but the British boatyards were occupied in building trawlers or repair work and none had the facilities to build in the quantities needed. In February 1915 the Admiralty started negotiations with the Electric Boat Company (Elco) of Bayonne, New Jersey, to construct fifty 22·9m (75ft) fast motor boats for use as submarine chasers and on 9 April a contract was signed. However, because America was not at war it was politic to build the boats in Canada and Henry Sutphen of Elco found and leased a small boatyard in Quebec with the promise of delivering the boats in one year's time. This of course could only be done on a prefabrication basis. Timbers for the boats were cut and shaped in Bayonne and delivered by train to Quebec for assembly: frames, knees, keels, decks, wiring harnesses, fittings, machinery were all sent to Canada. Yarrow had prefabricated many of his shallow-draught steamers which were needed in Africa, but they were supplied in ones and twos. Elco's noteworthy achievement was rigorous standardisation and superb organisation, intensified when the Admiralty requested an additional 500 boats. This meant that Elco had to find bigger premises, and it leased the yard of Canadian Vickers Ltd at Montreal and enlarged the one at Quebec. The second contract, signed in July 1915, was for 24·4m (80ft) boats of 37·6 tonnes (37 tons) displacement. The engines for all the boats were 164·1kW (220bhp) Standard motors, two in each boat. The increase in numbers was beyond some suppliers and Tiffany's studio, more at home in ornamental art, was successfully able to produce extra bronze castings for parts of the stern gear.

Fifty boats were delivered to the St Lawrence River before winter and another eighty-four went by train to Halifax from Quebec. The remainder were shipped to Halifax in the spring of 1916, where each boat was tested prior to acceptance by Admiralty representatives. They were safely shipped to England, fitted out and commissioned and worked up in Portsmouth before being sent out to Yacht Patrol bases in flotillas of two to ten boats as they were ready. The first three motor launches were in commission at Portsmouth on 14 October 1916. As the MLs entered service, so the original motor boats were paid off, to the mixed reaction of the RNVR crews. Undoubtedly many of the private motor boats had been comfortable, even luxurious craft. The MLs, though bigger, were spartan and the quarters forward for the crew of eight were cramped though the boats' two officers had a small wardroom aft. The engine installation was amidships and forward of that was the chartroom. The wheelhouse was open at the back with canvas spray dodgers at the sides and on top. These were often removed in fine weather. The boats were quite fast and had a best speed of 18 knots, but with their narrow beam of 3·7m (12ft) and shallow draught of 1·5m (5ft) they were uncomfortable boats in a seaway, rolling severely and liable to broach in a following sea. Nor were they dry boats, having little sheer, and lookouts posted on the foredeck during night patrols ended their watches rather wet. Nevertheless, crews put to sea in very poor weather conditions in the expectation of depth-charging a submarine, or stayed on station, tracking a submarine with their hydrophone until reinforcements arrived or it was forced to the surface. For such eventualities MLs were equipped with a Lewis gun mounted on the foredeck, though some MLs had a 7·6cm (3in) anti-aircraft gun. Most boats also carried rifles with which crews would relieve the monotony of patrols by potting at mines.

One of 550 narrow-gutted, uncomfortable and lightly armed MLs of the Auxiliary Patrol, which were manned dauntlessly during the First World War. *Imperial War Museum*

The MLs and drifters of the Auxiliary Patrol were so successful at their duties that when the submarine offensive started in the Mediterranean, detachments of the boats were sent there to supplement the work of the Italian fast motor boats. Some boats were lost when a transport was torpedoed in 1916 and in July 1917 a flotilla of eight MLs left Portsmouth and crossed France through the inland waterways to Marseilles. If the boats' seakeeping qualities had been better then MLs could have made the passage across the Bay of Biscay on their own keels, but even the ardent courage of RNVR crews would not have been sufficient for that journey to be made in safety. However, they were able to maintain their speed in moderate weather conditions if the circumstances warranted the ensuing lively ride. Although the MLs were bought in a hurry, the Admiralty might have been better advised to equip the Auxiliary Patrol with more substantial vessels, such as the American 33·5m (110ft) submarine chasers, which were slower than the MLs but were able to cross to Europe under their own power. Nevertheless, by the end of the war only 29 MLs had been lost in action, many by fire, and 30 more were laid down in 1918. The boats culminated their steady months of achievement when in 1918 they played a crucial rôle in the raids on Zeebrugge and Ostend; their counterparts in this expedition were the smaller and faster coastal motor boats and it is important to study their development and purpose for a fuller understanding of the rôle of fast boats in war.

During the summer of 1915, three lieutenants stationed at Harwich, Bremner, Hampden and Anson, realised that small fast motor boats could be rigged up with a torpedo, cross the North Sea, and go over the defensive minefields to beard the German navy in Schillig Roads. The idea was well received (unlike a plan in 1915 by Sir Percy Scott for fast motor boats to ram submarines, which did not get beyond the experimental stage) and tenders for a suitable design were requested from various boatbuilders. Only Thornycroft thought that the requirements could be fulfilled, partly because the company's experience with *Miranda IV* led towards such a boat and partly because Thornycroft was already experimenting with ideas for a similar small torpedo boat. The Admiralty saw the boat as being carried aboard a mother ship to its destination and then lowered into the water to get to its target at high speed. This was the thinking the Admiralty had applied to its first torpedo boats forty-five years earlier, forgetting that the range of petrol engine boats was much greater than steam boats, but not the fact that small boats at sea are highly vulnerable. This meant that Thornycroft's design could not exceed the weight of the 9·1m (30ft) motor boats carried in cruisers' davits, 4·3 tonnes (4¼ tons), and had to be fitted with lifting gear. The boat had to be able to carry a crew of three, fuel for 200 miles, and a 0·76-tonne (¾-ton) torpedo at a speed of not less than 30 knots.

The design which the Admiralty approved was similar to *Miranda IV*, with a single step amidships, following the line of the hull, but with a marked chine from the step to the stem. The boat was 12·2m (40ft) long and 2·6m (8ft 6in) in beam. The problem of carrying the torpedo was solved by placing it in a trough and firing it astern. The Admiralty had vetoed the idea of firing the torpedo over the bows, although Italian motor torpedo boats did this with success. It still was not possible to use side-dropping gear without slowing the boat almost to a halt. However the objection was raised that to fire the torpedo over the stern the boat would have to turn stern to the enemy. To meet this, Thornycroft took advantage of the boat's high speed of attack to suggest firing the torpedo tail first, giving the boat time to veer away from the track of the torpedo.

Both Hampden and Bremner collaborated with Thornycroft during the design of the boat, known at first as a submarine vedette, and then during construction once the design had been accepted and an order for twelve boats placed. They were built in 1915 at Thornycroft's boatyard on Platt's Eyot on the Thames by Hampton under conditions of great secrecy, thereby attracting considerable attention. Until this time the Hampton yard had been building fast launches with Thornycroft engines for the services. Because of the tumblehome it was found easier to build the boats upside-down, only turning them over for completion. They were built with three skins of mahogany planking on closely spaced steamed ribs of American elm. The engine was mounted on deep bearers which ran the length of the boat and the step was built on to the hull, creating a space of 7·6 or 10·2cm (3 or 4in) which was usually packed with cork. The trough for the torpedo was an integral piece of the after part of the hull, supported by timbers attached to the bearers, and was itself made of wood though the supporting rails were bronze faced. The lifting slings were fastened to the bearers ahead of the engine and/or either side of the trough, though few boats were ever carried aboard ships. The torpedo was fired from a lever in the cockpit which detonated a cordite charge, an aluminium-capped ram pushing the torpedo down the trough. The 45·7cm (18in) Mark VIII was the standard weapon, set for 1371·6m (1500yd) at 41 knots or 2275cm (3500yd) at 29 knots. The torpedo could be locked in place during heavy weather. The cockpit, just large enough for the two officers, was encased between the engine forward and the brass 454-litre (100-gallon) fuel tank aft. The chart table and compass were mounted in front of the helmsman. The luckless mechanic had a seat below deck to starboard, next to the bilge pump, and usually spent most of his voyage tending the engine. In the first boats they were Thornycroft 186·5kW (250bhp) V12 units, with only neutral and ahead gearboxes. To keep weight down the boats were sparsely fitted out; it was correctly presumed that the keen men who would be drawn towards service in these boats could manage without such luxuries as an enclosed wheelhouse. On covert trials conducted on the Thames Estuary the boats were found to be capable of 33½ knots with a full load. This was a commendable speed and the boats were well regarded by all concerned. The first boat was launched on 6 April 1916.

The question of where to station the flotilla was thought to have been solved when Lieutenant Hampden proposed a secret base on Osea Island in the Blackwater River, Essex. Plans were advanced to set up an establishment there when a counter-suggestion was made recommending Queenborough in the Swale, already a submarine depot. It was important that the boats were kept in good shape and it was essential that they could be lifted out of the water at the end of a patrol, otherwise the timbers absorbed water and prolonged immersion caused galvanic action between the propellor shaft and the bronze propellor. J. E. Thornycroft stressed this matter in a letter to Hampden at HMS *Vernon:*

'I think you yourself appreciate the necessity, but the authorities to whom this type of boat is new can hardly understand, without having had personal experience, how, first of all, to get the speed performance everything must be in the very best condition, and to maintain their speed and seaworthiness every detail must be looked after.'[3]

The man responsible for the boats' organisation at Queenborough was Captain Barry Domville. He thought the conception of the skimmers was a 'thoroughly sound offensive operation of a minor order'.[4] It was continually stressed that the boats were for offensive use and that any diminution of their rôle would seriously handicap their capabilities. Unfortunately this is what happened. Thornycroft produced the best boat that could be expected in the time and within the prescribed limits, and valiant officers were forthcoming to man the craft, but little account was taken of the difficulties of fighting in the North Sea in a sprightly 12·2m (40ft) boat. Accurate navigation was out of the question, even when the boats were fitted with two compasses, and dead reckoning had to be worked out from the engine revolutions. Even then, one part of the flat, occupied North European coast looked much like another, and Domville sensibly coopted Erskine Childers who knew the coast from personal experience. Thornycroft's engines were far from satisfactory: pipe unions suffered from the pounding at sea and the long cylinder blocks were liable to fracture. The boats were very wet when running at slow speeds into a seaway, but most disappointing of all was their poor performance in fresh winds. The Admiralty was moderately protective towards the boats and did not permit their first raid to be carried out without air reconnaissance. In October 1916 permission was finally received for an operation against German vessels lying in Schillig Roads. The boats lay off the German coast and two seaplanes from the carrier *Vindex* actually landed in the Roads but reported that heavy coastal fog made any attack by small boats inadvisable. In fact conditions were ideal for a stealthy attack, although the boats' noisy exhausts would probably have pinpointed their position. They had two Lewis guns on a single mounting. After this sortie the boats were committed to reconnaissance work inside minefields and came to be officially known as Coastal Motor Boats. A detachment of CMBs was stationed at the Dover Patrol's advance base at Dunkirk, assisting in submarine hunting and awaiting the opportunity to show their paces. Their inaction was not unobserved. In February 1917 the Vice-Admiral at Dover, R. H. Bacon, wrote to Dunkirk:

> 'In new designs such as coastal motor boats the officers should use their brains to overcome difficulties, otherwise the boats will never be of use. They should buckle to, and make the boats a working war success, and not wait for others to do it for them. At present the coastal motor boats have chiefly come under my notice for things they cannot do; I wish soon to see that altered into things they can do.'[5]

The message was understood and in March a pack of four CMBs attacked some German destroyers off Ostend: the large new destroyer G 88 was sunk and others were damaged. This success catalysed further interest and orders for five 16·8m (55ft) CMBs, for which Thornycroft had built a prototype, were put in hand. These were twin-screw vessels of 8·1 tonnes (8 tons). They were better sea-boats and could maintain a higher speed in a seaway than the 40-footers, which could not manage more than 15 knots in sea state 4. The 55-foot boats had the advantage of more reliable engines but these could not be produced in sufficient quantities and the building programme was seriously delayed. The first five had the same engines as the 12·2m (40ft) type which only gave a speed of 32 knots but subsequently Sunbeam and Green aviation engines were installed as they became available: some boats had twin 186·5kW (250bhp) Greens and a few others twin 335·7kW (450bhp) Greens and twin 335·7kW (450bhp) Sunbeams. By 1918 Thornycroft had produced a 279·8kW (375bhp) version of its V12 and with the higher power engines the 16·8m (55ft) boats could reach speeds of 40 knots. The boats had two troughs and carried any permutation of two torpedoes and four depth charges. However, at much less than maximum speed, the boats could not safely fire their torpedoes, which further reduced their effectiveness in fresh conditions. In form and construction they were substantially the same as the smaller boats, though built with heavier scantlings and incorporating a small deckhouse. Most of the boats

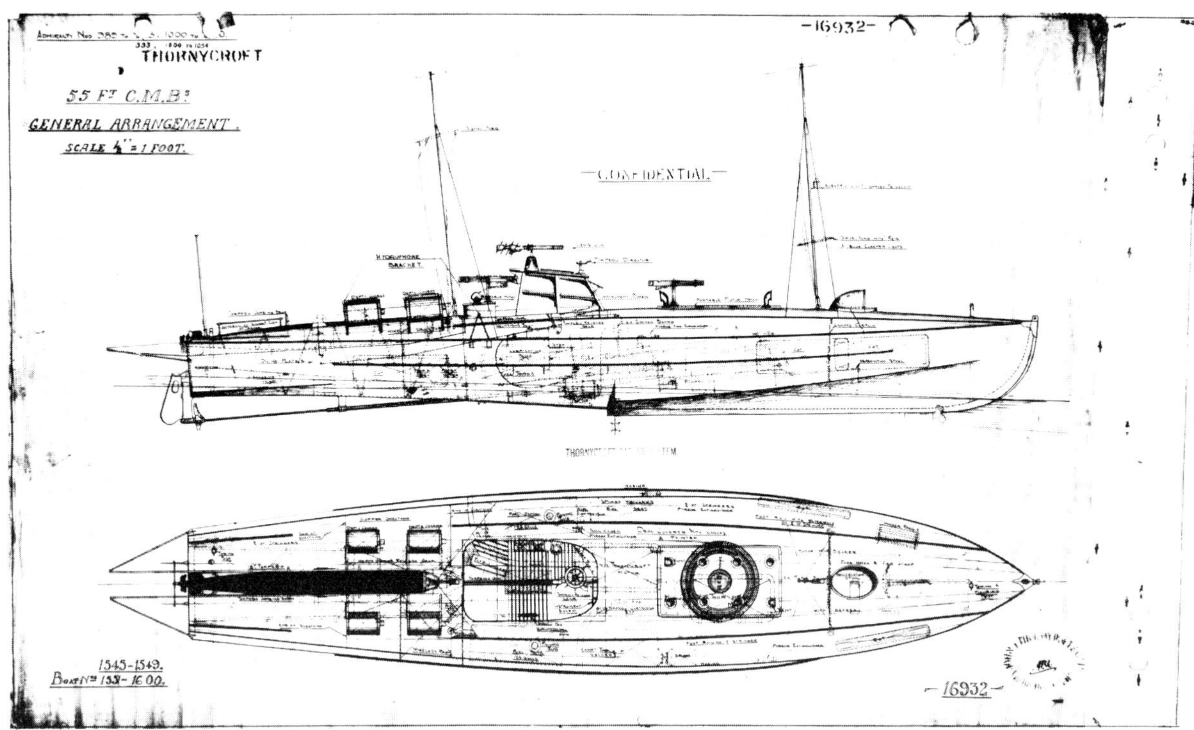

had wireless and usually carried a W/T operator and an additional mechanic. Their fuel capacity was 1589 litres (350 gallons) which gave them a range of 200 miles and they mounted four Lewis guns. Fifty-six of these CMBs were in service by 1918: twelve at Harwich, twenty at Portland, the main Channel base for anti-submarine operations, and twenty-four at Dover and Dunkirk. Most 16·8m (55ft) CMBs were capable of 35 knots and the higher powered boats could reach over 40 knots, but rarely did the occasion, or the sea conditions, permit high speeds. Most of the time the boats were minelaying off the Belgian coast (fifty-five mines were laid close to Zeebrugge and Ostend in 1917 and 1918), helping to escort the French coal trade (Portsmouth-Havre) and supporting monitors off Belgium. Some boats were rigged with directional hydrophones for anti-submarine patrols, a task for which they were hopelessly unsuited. It was only in quiet conditions that a boat's wireless was operational, because at planing speeds noise and interference from the engines was considerable and a wireless operator was not always carried. Some of the boats were fitted with spiral propellers which did not foul on defensive obstructions such as nets. With their shallow draught the boats were able to carry out valuable reconnaissance work inshore of minefields, where in the sheltered waters they could use their high speeds to considerable advantage. Their profile was very low; even if detected in daylight raids by shore batteries, they could cover their tracks with smoke screens.

Largely as a result of a successful attack off Ostend in 1917 by four 12.2m (40ft) CMBs led by Lieutenant W. N. T. Beckett *(CMB 4)*, in which a German destroyer was sunk and others damaged, it was decided to order further boats, including the 16.8m (55ft) type. The plans are of Thornycroft's prototype, in which four depth charges were carried for anti-submarine work; later boats were fitted with two torpedoes.

The German submarine menace was not finally contained until the convoy system was established in the second half of 1917. Compared with the value of convoy protection, the anti-submarine measures of the Auxiliary Patrol were minimal, but CMBs were held in high regard by Germany and some effort was directed towards finding similarly useful boats. In August 1915 the German navy renewed experiments with remote-controlled boats and ordered twelve from Lürssen. The first nine were for use against British monitors bombarding the Belgian coast. They were timber boats with twin 6-cylinder Maybach diesels totalling 313·3kW (420bhp), giving a top speed of 30 knots. Known as

Fernlenkboote (remote control boats), they were loaded with 681·0kg (1500lb) of explosive and controlled by an insulated cable run out from a drum in the stern; a spotter plane relayed directions to a shore-based operator. One of these boats, apparently out of control, blew up part of a pier at Nieuport in March 1917, and an ML sank the first FL raider with gunfire. But in October 1916 FL12 had struck the monitor *Erebrus* amidships and caused some damage. Another Lürssen boat, *Boncourt,* 9·5m (31ft), was built in 1913 and commissioned in 1916 to sabotage British mine nets off Belgium. She failed in this and was equipped with a torpedo tube in her bows and used in the Baltic, where on 10 September 1917 she sank a Russian steamer. Being so small, *Boncourt* was carried aboard a parent ship but she was a hard chine boat and could reach 43 knots in good conditions.

Germany did not have the same requirements for coastal boats as Britain, needing primarily motor patrol boats, and seventy-five 16·5m (54ft) boats of 10-11 knots were built in 1915. They were fitted with two petrol engines and carried one torpedo on deck and six mines. The mine nets were a hazard to German submarines and in the summer of 1916 an order was placed for six fast motor torpedo boats which could destroy the nets, and by 1918 twenty of these boats had been commissioned. The only engines available were spare 6-cylinder Maybach petrol engines from naval Zeppelins *(Luftschiff–Motoren)* and the boats were known as the LM class. They had three engines each, producing between 470kW (630bhp) and 537kW (720bhp) and this gave a speed of 28-32 knots. The boats were between 14·6m and 17·1m (48 and 56ft) in length. They had one 45cm (17·7in) torpedo tube in the bow and a 31mm (1·5in) gun. The German Navy demanded good performance in a variety of sea conditions when placing tenders for boats, and the LM boats were expected to be able to maintain 29 knots in sea state 3 and not have to return to base until at least a Force 5 wind was blowing. Accordingly the boats, most built by Lürssen, others by Max Oertz of Hamburg and Roland-Werft Vertens of Bremen, were round bilge displacement craft, starting a tradition in design which has characterised German fast boats to the present day. Arising from the LM type was a superior class of boat, ordered in March 1918 but delivered after the war. They were 18·9m (62ft) boats with three petrol engines and mounting two torpedo tubes on the foredeck, and were classified by their engines and builders; thus Juno (261·1kW [350bhp] Junkers, four boats built by Oertz); Koro (261·1kW [350bhp] Korting engines, two boats built by Roland) and Lusi (298·4kW [400bhp] Siemens, two boats built by Lürssen). All the boats carried two 20mm cannon and the Koro boats could reach 41 knots. In view of the limitations imposed on German naval construction by the Treaty of Versailles, these foundling boats in the German fast boat arm, the *Schnellbootswaffe,* were instrumental in shaping German thinking when in 1924 they were used by the navy for full evaluation.

As early as November 1916 it was realised that one way to tackle the German submarines was to bottle them up in their base, which was in Bruges. In concrete pens they were immune to the trifling bombs of the Royal Flying Corps, and had access to the coastal ports of Ostend and Zeebrugge by two short canals. There had been an abortive land offensive in 1917 and CMB raids off the coastal ports which at least denied Germany unrestricted use of the two ports; two torpedo boats were torpedoed off Zeebrugge and one off Ostend, and at least one ship was sunk by a mine. But little was done until the beginning of 1918 when Admiral Keyes, taking over from Bacon at Dover, proposed blocking the ports with hulks. It was an important and well planned raid, involving 166 vessels most of which were Auxiliary Patrol craft, 24 CMBs, and

61 MLs, which were at last able to demonstrate capabilities which for so long had been subjugated to the routine of patrol responsibilities. Their rôle was small but crucial (laying navigation marks and flares and destroying a pierhead) and was as important a contribution to Auxiliary Patrol morale as was news of the raid itself to the flagging spirits of Allied troops on the Western Front.

The force left the Thames Estuary in the afternoon of 22 April. The CMBs were towed across, as their fuel consumption would have been abnormally high if they had gone at the speed of the rest of the force. The force divided twelve miles from the shore, one division going south to attack Ostend and one continuing to Zeebrugge. At 11.40pm MLs and CMBs started laying smoke screens inshore at both harbours. At Ostend two groups of six MLs screened areas to the east and west of the port and the central section was covered by the faster CMBs, all the boats being covered by eight destroyers standing out to sea. However, because the Germans had shifted a buoy marking the line of entrance to the harbour the two blockships, the cruisers *Brilliant* and *Sirius*, piled harmlessly on the beach and their crews were taken off by MLs. At Zeebrugge there were three blockships, the cruisers *Intrepid, Iphigenia* and *Thetis*, each packed with 1524 tonnes (1500 tons) of concrete, and under the smokescreens provided by the MLs and CMBs the first two were able to enter the harbour mole and scuttle themselves right in the entrance of the canal. Three other ships landed at the seaward side of the mole and knocked out the battery at the end of it, and a submarine was exploded beneath the causeway which connected the mole to land, thus preventing German reinforcements from reaching the battery. The raiders encountered negligible opposition from

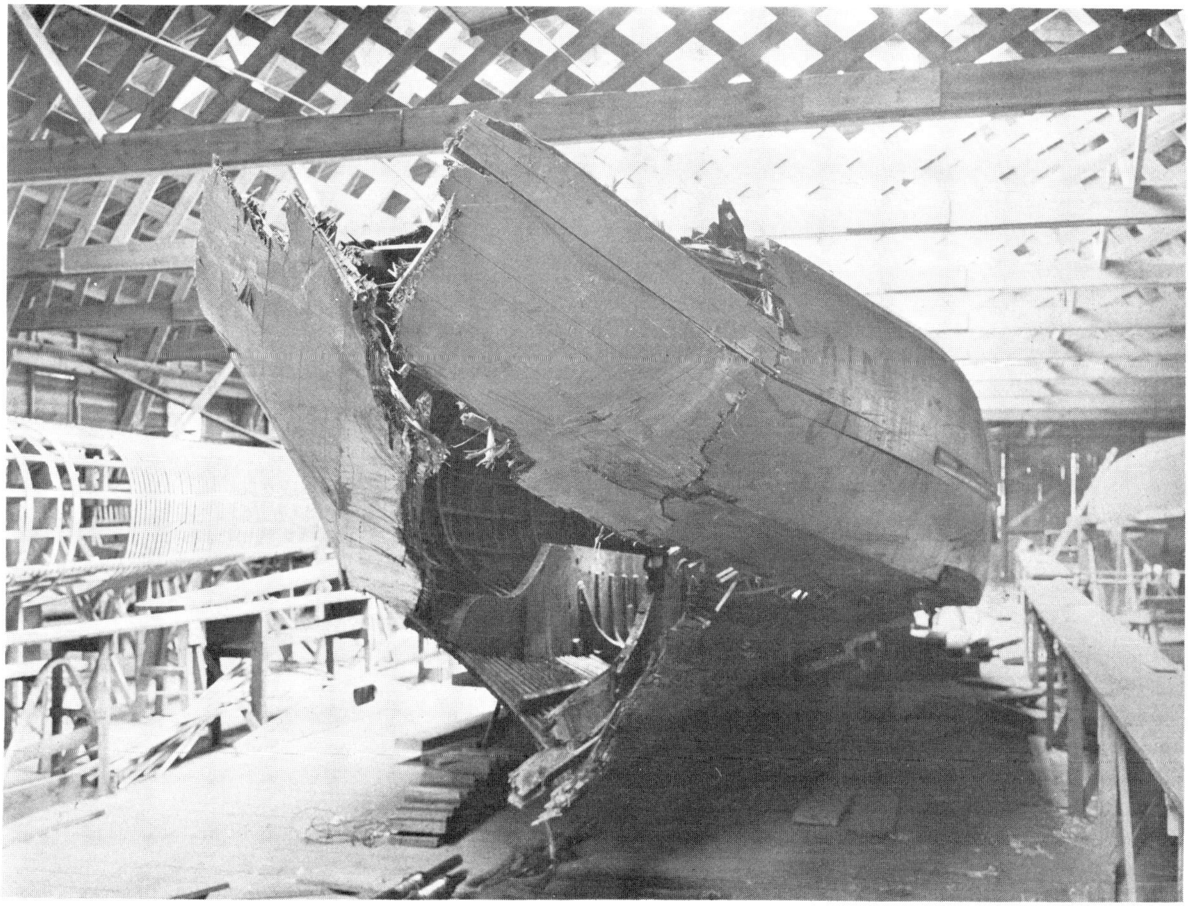

An attempt to sink a submarine by ramming caused the damage to this CMB. She is being repaired at Thornycroft's Thames yard at Hampton, where another CMB is having ribs of pliable American elm fastened around the hull. A dozen other yards also built CMBs for the Royal Navy during the First World War and a total of 123 was produced.

German destroyers or patrol boats as none seemed to be on duty that night. Only one destroyer emerged from the docks and she was struck by a torpedo from a CMB. The MLs took off the crews from the scuttled ships under intense fire from shore batteries. *Thetis,* which had cleared the net obstruction when entering the harbour, was forced to stop short of the canal with her propellers fouled, but an ML *(522)* took off her crew and the crew of *Intrepid* in ten minutes. Two MLs were sunk and there were casualties aboard other boats but the raid was considered sufficiently successful for another attempt to be made on Ostend on 9-10 May. *Vindictive,* which had played a crucial part in the Zeebrugge attack, was the blockship, and guided in by a flare sent up close inshore by a CMB. This raid was made from Dunkirk, without any preliminary bombardment, and the MLs and CMBs made their smoke screens in stealth just off the coast. An ML *(254)* took off thirty-nine men from the scuttled *Vindictive* but was herself so badly damaged by machine gun fire that everyone was transferred to the destroyer *Warwick* before the ML sank. Another ML *(276)* rescued some men from the water and escaped, though badly shot up. As in the previous raids opposition from German surface craft was light, though one CMB *(22)* attracted the searchlight and then the guns of a torpedo boat lying close inshore. As this boat and other CMBs had to fire their torpedoes at the harbour piers, she attacked the torpedo boat with such vigorous fire from her Lewis guns that the torpedo boat was driven away from the harbour.

The raids were successful exercises, and a boost to morale, but none was successful for long. By 14 May a channel had been cleared around the blockships at Zeebrugge and destroyers and submarines were operating out of Bruges as before, and *Vindictive* was soon hauled out of the entrance to the Ostend canal. The triumphant rôle of the men and boats of Auxiliary Patrol in the operations was widely recognised; they had behaved with discipline, bravery and initiative, and the Victoria Cross which was awarded to Lieutenant P. T. Dean of *ML 282* for his rescue of the men from *Iphigenia* was emotionally shared by the RNVR as a commendation of its officers and men.

The sheltered waters of the Mediterranean led the Royal Italian Navy to experiment with torpedo-carrying motor boats on a serious scale as early as 1914. As in England there had been some toying with the possibilities of such boats since 1906, and in the Dalmatian coast lay an equivalent hunting ground to the Frisian Islands. Enthusiastic motor-yachtsmen had formed the *Corpo Nazionale Volontari Motonauti* in the summer of 1915. Using their own boats, they patrolled coasts and harbour approaches until the first specially designed torpedo craft entered service. There was even an Italian Thornycroft, the Venetian yard of Societa Veneziana Automobili Nautiche, known as SVAN, the biggest of many Italian boatyards which had been building motor boats since the introduction of the petrol engine. The first two experimental torpedo boats were laid down and launched in 1915. They were hard chine timber boats, 15·9m (52ft) long, with two 167·9kW (225bhp) Isotta Fraschini petrol engines which gave speed of 23 knots. These were 6-cylinder aviation units, developed for use in Italian airships and had a remarkably low power/weight ratio of 3lb per bhp. They evolved into the similarly excellent V8 engines used in some British and most Italian boats before and during the Second World War. At first the boats fired their two torpedoes from internal tubes aft but subsequent boats nearly all carried their torpedoes in dropping gear on the sidedeck or suspended outboard. The boats were known as MAS: *Motobarca armata* SVAN (SVAN-built armed motor-boats) though the acronym stood for different things as the functions of the craft changed, ending up as *Motoscafo anti-sommergibile* (anti-submarine motor-boats). During 1916 two more series of MAS-boats were built by SVAN, the first nineteen 24-knot boats having twin Isotta Fraschini engines, the next twenty-five having twin 149·2kW (200bhp) Sterlings, which gave them only a maximum speed of 21 knots or so.

Most of the MAS-boats were built by SVAN but towards 1918 other yards such as Piaggio and Gallinari built SVAN-type boats, while Orlando of Livorno and Baglietto of Varazze built their own types of MAS. Large numbers of motor launches were also ordered from the Electric Boat Company of New Jersey, still classed as MAS-boats. They were rather larger than the Italian MAS-boats and were fitted with twin Standard petrol engines, and though slower they had a greater range than the Italian boats. Some later Elcos were built by Italian yards.

Most MAS-boats were between 15·2 and 22·9m (50 and 75ft) and capable of 24-26 knots, though Baglietto and SVAN built some faster types of 30-33 knots, which were not completed until after the war. Nevertheless, they acquired a reputation among the Austrians for considerable speed. One of the most useful features on Baglietto and Orlando boats was the pair of 3·7kW (5hp) Rognini electric motors which gave them a speed of 4 knots for up to four hours. Of all small offensive craft during the war, only MAS-boats were used consistently as they were designed to be: in local, stealthy raids on enemy shipping and installations in harbours. MAS-boats had large crews, usually eight, and they were trained in special courses established for the service. They were organised in flotillas of fifty to sixty and they normally hunted in pairs. Their first successful venture was on the night of 7 June 1916, when *MAS 5* and *MAS 7*, which were not fitted with electric motors, torpedoed an Austrian steamer after getting through the net defences of Durazzo. On 26 June the same pair of boats again penetrated Durazzo and sank two steamers, but there were no more major successes until the following year, during which time the MAS organisation was consolidated and many more boats were acquired.

Towards the end of the war five 21.3m (70ft) CMBs were ordered for minelaying duties. They were fitted with a variety of engines and were capable of 26 knots. They were probably the largest hydroplanes ever built but they were not delivered until 1919. *Peter Garnier*

The Austrian navy had grown wise to MAS-boat tactics and sneak raids became more difficult. An MAS-boat entered the anchorage at Fasana on 2 November 1916 (the barrage was forced down with weights by an accompanying torpedo boat) and spent two hours searching for the battleship *Ferdinand Max*. Unable to find her, the MAS-boat fired two torpedoes at an old battleship, *Mars,* at 182m (200yd) but the ship's boom nets entangled the torpedoes. However the MAS did manage to escape from the harbour undetected. The next encounter was a daylight one, when the Austrian battleships *Wien* and *Budapest* and a handful of torpedo boats steamed out of Trieste on 16 November 1917 to bombard Cortallezzo. Two MAS-boats left Venice to intercept them and, under cover of a sea mist, closed to 1400m (1500yd) before attracting fire from *Wien.* The boats closed to 750m (800yd) before firing their torpedoes and turning away, when they were pursued by torpedo boats. The stem of one MAS-boat was shot away but its high-speed planing angle prevented water from filling the boat. Both boats' torpedoes missed but the attack was sufficient for the battleships to terminate their bombardment and return to Trieste. The Austrian navy were, of course, wary of MAS-boats and having had little success in preventing this attack, the decision to retire was prudent. However, it is probable that further attacks could have been parried by the alerted torpedo boats if more officers had decided to tackle the Austrians, and it was reprehensible that the MAS-boats' torpedoes, fired in a calm sea, failed to hit the ships. Following this spirited sally, two MAS-boats caught the Austrian navy in harbour in a classic fast fighting boat raid. Escorted by torpedo boats, on the night of 9-10 December 1917 *MAS 9* and *MAS 13* left Venice for Trieste, making the last five miles under their electric motors. Men were landed at a mole, to spend two hours cutting through barrage cables, and another twenty-five minutes' steaming brought *MAS 13* to within 200m (200yd) of *Wien.* Astern of her and further inshore was *Budapest* but *MAS 13* could not manouevre into a suitable firing position. The boats fired their torpedoes together and *MAS 9*'s struck *Wien* amidships and she quickly sank. Both torpedoes of *MAS 13* missed *Budapest* and, though she was under fire for five minutes while trying to start her main engines, she succeeded in escaping without damage and both boats returned to Venice.

For a raid on the heavily defended harbour of Pola the Italian navy built four punt-like boats in early 1918, which were able to crawl over barrages by means of caterpillar chains. Two of the boats had to be scuttled off Pola in an unsuccessful attack on 13 April but on 14 May the 16m (52ft) *Grillo* did clamber over three barrages. She was spotted immediately and came under fire from a battleship, and the crew were taken prisoner by an armed steam boat.

On 10 June 1918 the Austrian battle division was steaming for Cattaro to destroy with mines the barrage at Otranto. That morning Lieutenant Luigi Rizzo, who had sunk *Wien,* was finishing his patrol when he saw smoke on the horizon and thought that Austrian destroyers were making for his patrol of two boats, *MAS 15* and *MAS 21.* He decided to attack and found himself off Premuda Island in a sub-division of two 20,000·0-tonne (20,000-ton) super-dreadnoughts, *Tegetthoff* and *Szent Istvan,* and ten destroyers. Remarkably, the MAS-boats were undetected in the half-light of early morning, and in *MAS 15* Rizzo penetrated the destroyer screen on *Szent Istvan*'s starboard flank at 12 knots and fired two 45cm (17·7in) torpedoes at the battleship. They struck amidships and she went down after about two hours; her last moments were filmed in one of the most spectacular recorded sequences of the First World War. The other boat, *MAS 21,* failed to strike *Tegetthoff* and the pair cleared away at their best speed of 24 knots. They were chased by a destroyer but Rizzo dumped two depth charges in her path, one of which exploded under her bow and she abandoned the chase. Both boats escaped.[6]

Other MAS-boats achieved successes against transports along the Adriatic coast and frequently served as minelayers and scouting craft. The Austro-Hungarian navy made little effort to counter the Italian boats with small fast craft of its own, although in 1915 *Linienschiffsleutnant* von Müller-Thomamühl designed the world's first surface effect warship. She was 16m (52ft 6in) long and powered by four 22·4kW (30bhp) Austro-Daimler aero engines driving two water-propellers. A 48·5kW (65bhp) Austro-Daimler engine drove a centrifugal fan which provided the vertical lift; the boat was capable of 32·6 knots in calm conditions. She carried two 35cm (13·8in) torpedoes launched nose-first over the stern but did not prove seaworthy enough for service in the Adriatic.

The Italian MAS-boat force, which numbered over 350 by the end of the war, realised the full potential of torpedo motor boats, but in comparing their rate of success with CMBs it should be remembered that their targets were more readily apprehended; CMBs were not suited to patrolling and so were less likely to make chance encounters with German ships, and the disposition of the High Seas fleet made strikes in port difficult to achieve. Moreover, MAS were given every encouragement by the naval staff. Notwithstanding their good reputation with the Auxiliary Patrol, CMBs failed to perform many of the feats expected of them, and they were not regarded with much favour by the Admiralty.

By 1922, 422 MAS-boats had been commissioned, nearly all of them built to the same requirement of small size and, apart from the Elco boats, none of the Italian craft was larger than 22m (72ft) in length. The Italian navy was the only one which resisted the inherent inclination to increase the size, seaworthiness and armament of its small boats, realising that the zeal and licence of crews declined as the size and immunity of the boats was increased. There were very few active service losses in the MAS service. Most boats were lost through petrol fires started in docks, during or shortly after refuelling. Nearly all the wartime boats were discarded during the 1920s but strong interest was maintained by the Italian navy and a handful of prototypes and improved versions were built in the 1920s and 1930s, providing the navy with strategies and the trained men to execute them

5 MORATORIUM

After the war the Admiralty was glad of the opportunity to run down the fleet of auxiliary motor boats. The MLs were sold, to be converted into houseboats or motor yachts; some still provide accommodation in south coast mud-berths. War losses of CMBs had been light and to keep the flotillas up to strength a few new boats were ordered from Thornycroft, including five 21·3m (70ft) boats of 28·5 knots for minelaying. Two had twin 261·1kW (350bhp) Thornycroft engines, two had 373kW (500bhp) Greens, and a fifth, much faster boat had a pair of 24-cylinder Thornycroft engines. But despite some excellent work in the Baltic in 1919, the CMBs were never to receive wholehearted favour.

In December 1918 a flotilla of 12·2m (40ft) CMBs was sent to the Caspian to lend support against Bolshevik gunboats on the Volga estuary and in August 1919 a detachment of 16·5m (55ft) boats covered the rearguard of the British withdrawal at Archangel. Other CMBs were used, with many more MLs, patrolling Rhine bridgeheads with the British Army of occupation. The undertaking in the Baltic was a secret service operation of classic form. A British agent in Russia (Sir Paul Dukes) was isolated in Leningrad because of the capture of his couriers. It was proposed that a CMB, operating from a Finnish harbour three miles from the Russian border, could slip past the immense island fortress of Kronstadt and rendezvous with the agent. This operation was entrusted to Lieutenant A.W.S. Agar, an experienced CMB commander. He supervised the preparation of two 12·2m (40ft) boats, which were then shipped to Sweden, then Helsingfors and then towed by destroyer to Bjorko, from where they made their own way. The whole affair was a delicate one, calling for tact, courage, and seamanship of the highest order. In *CMB 4* (the spare boat was *CMB 7,* Lieutenant J. Sindall) Agar sidled past the fort's defences night after night and at dead slow speeds, opening up to 30 knots and then standing off the Russian coast to make contact with the agent. There were many harrowing occasions but due to the vigilance of the crews and near-faultless behaviour of the boats, the operation was largely successful.[1]

During this furtive work Agar made a night attack on the Bolshevik cruiser *Oleg,* creeping through an alley of destroyers and firing his torpedo 450m (500yd) from the vessel, just as he was spotted and fired upon. The torpedo struck *Oleg,* which shortly sank, and Agar escaped unscathed. The merits of the CMBs attracted the attention of Admiral Sir Walter Cowan, of the Baltic peace-keeping fleet, and he decided to use the boats in a raid on Kronstadt to immobilise the Russian capital ships in the harbour, and so free his force to cope with recalcitrant Germans at the western end of the Baltic. Meticulously he planned a raid, with air support, for 18 August 1919. Aircraft were to provide a diversion while cruisers and destroyers were stationed outside the minefield which defended the harbour. Six 16·5m (55ft) CMBs *(24, 31, 62, 72, 79, 88)* were to penetrate the minefield and make torpedo attacks inside the harbour; a seventh boat, Agar's, was to stand off outside the harbour to handle any patrol craft which might emerge.

As Sopwith Camels and Short seaplanes dropped little bombs and drew attention away from the surface, the first three boats roared into the congested harbour. Lieutenant Bremner *(79)* fired his torpedo and sank a submarine depot ship; Commander C. Dobson *(31)* struck the battleship *Andrei Pervosvanni* with two torpedoes and Lieutenant A. Dayrell-Reed *(88)* struck the battleship *Petropavlosk*. By this time the harbour was in pandemonium, the advantage of surprise had gone and the remaining boats failed to achieve anything but add to the chaos and rescue the casualties. In the mêlée three boats *(24, 62, 79)* were lost and eight men were killed. The two battleships were so badly damaged that they remained in the harbour for months undergoing repairs and were unable to leave before winter ice locked them in. This raid on one of the best defended fortresses in the world was an outstanding success and an indication of the value of suitable small craft when used imaginatively in organised and well-controlled operations. (See Agar for full account.) But at an Admiralty conference held in May 1920 to decide the future of the CMB it was felt that:

'In view of the present political situation and the large reduction anticipated [correctly] in future Naval estimates, and taking into consideration the short life of CMBs, it is not considered advisable for the present to allocate CMBs to local defence or the main fleet.'[2]

A 12.2m (40ft) CMB, part of the Allied campaign against the Bolsheviks, at work in the Caspian in 1919. Other boats were sent to Archangel and the Baltic. *Imperial War Museum*

However, the base which had been set up at Osea Island in 1918 was retained to provide a skeletal centre for CMB training, minor operations and such experimental work as might transpire from new developments. This last aspect was a sore point with the Admiralty, which – harried by the CMB lobby – had never been entirely satisfied with the overall performance of the hydroplanes and was even being pressed to look at the designs of other boatbuilders. At a time when the German navy was building up its experience with fast motor boats because of the restrictions of the Treaty of Versailles, the Royal Navy was obliged by costs, the Washington Treaty of 1921-2 and the London Treaty of 1924, to limit naval building. Within those restrictions the Navy built ships at the top end of its requirements. Not only were CMBs discarded from 1921 onwards, they were not replaced, and in 1926 the CMB branch was paid off. This was not surprising in view of their relegation to coastal defence and patrol duties, for which they were hopelessly unsuited. For Thornycroft, the design continued to be a success for many CMBs were sold to foreign navies. As in the 1880s, lesser powers were keen to purchase torpedo boats from England, while, as before, the Royal Navy understandably lost its interest once hostilities ceased. In 1921 the French navy bought two CMBs from Thornycroft, one 13·7m (45ft) and one 16·8m (55ft) type, and thereafter the firm sold a further 27 boats before 1930, including two to the US Navy for evaluation in 1922, and others to Spain, Sweden, Siam, Japan, Finland, Netherlands, Greece and Yugoslavia. Between 1930 and 1938 fifteen were sold – most of them to China. Most of these sales were of 16·8m (55ft) types, and they were all fitted with Thornycroft engines. By 1936 the designs had been worked up to achieve 45 knots, though not all the boats were armed with torpedoes, and some were used as fast picket boats.

During the 1920s there was considerable interest in flying boats and the Admiralty bestowed on the RAF what it thought was necessary in the way of tenders and small craft to service the flying boats. Naturally these were similar to the Navy's own small boats and it was not until the end of the decade that the RAF Marine Service was formed and began to place orders for boats more suited to its needs. Naturally Thornycroft was a key supplier but it is noteworthy that, though accustomed to stepped hulls on flying boats, the RAF's first fast launches were 18·9m (62ft) hard-chine planing boats of 26 knots. They were built by Thornycroft, with twin 279·8kW (375bhp) Thornycroft petrol engines, and the first was delivered in 1927.

It was becoming apparent to many people that while the hydroplane was very good, it was not good for very much. A handful of CMB hulls had been fitted out as fast motor yachts after the war, but on the whole the hydroplane became restricted to BI Trophy competitors and outboard powered runabouts; the coming design was the hard-chine, stepless hull of concave-V sections forward and flattish sections aft. However, in the austerity of the 1920s the outboard engine became quite popular and enthusiasts frequently raced their hydroplanes, though not to the same extent as in America, where offshore as well as circuit races were all the rage. Designers and constructors of boats for the BI Trophy learnt a great deal about the performance of fast craft under stress. Before the war the successful racing boats were usually those most able to keep going. From 1920, when America challenged for the BI Trophy, systems were rather more reliable and victory was for the better designed and engineered boats.

In 1920 a challenge for the BI Trophy was issued by Commodore Garfield Wood, Ford agent turned industrialist, with his profitable invention of an hydraulic lift for Model T trucks. Wood was a keen hydroplane competitor, acquiring his boats from the boatyard of C. C. Smith & Sons. Chris Smith's yard at Algonac, Detroit, habitually turned out highly successful boats, such as *Miss Reliance II,* a defender in the 1912 BI Trophy races, *Miss Minneapolis* and *Miss Margaret.* Gar Wood learnt a great deal from Smith (who went on to found the prosperous Chris Craft corporation) and started to build boats to his own ideas; he drubbed all opposition to win the Gold Cup Trophy five years in succession, until in 1922 the American Powerboat Association ended the Wood monopoly by ruling hydroplanes out of the competition. Much of Wood's success came from abandoning marine petrol engines and fitting aircraft engines, which then offered the best power/weight ratios available. In 1918 he modified a Curtiss V12 and installed it in *Miss Detroit III,* claiming 60 knots in his Gold Cup victory and making the use of aviation engines popular among the swells that could afford them for racing and record-breaking craft. Wood challenged for the BI Trophy with two boats, *Miss Detroit V* and *Miss America I,* a 7·9m (26ft) hydroplane powered by two 373·0kW (500bhp) Liberty engines, developed by the Packard Motor Car Company for aircraft, and modified by Wood. Defending the trophy in 1920 were Mackay Edgar's *Maple Leaf V,* powered by four Sunbeam engines, *Maple Leaf VI,* with two Rolls-Royce engines and driven by a First World War pilot Harry G. Hawker, and a boat called *Sunbeam-Despujols.* The races were held on the Solent; both Wood's boats outclassed the defenders and *Miss America I* won the Trophy at an average speed of 53·4 knots. The following year Mackay Edgar challenged Wood for the Trophy with a new boat. *Maple Leaf VII* was a single-stepped 10·2m (33ft 6in) hydroplane powered by four 335·7kW (450bhp) Sunbeams, built by Saunders of sewn mahogany laminates, and capable of 69·5 knots. She was slightly faster than Wood's new defender, *Miss America II,* a 9·8m (32ft) hydroplane with four 373kW (500bhp) Liberty engines, but she punched a hole in her hull on the second lap of the race and once again the rest of the event was between competing American boats; *Miss America II* won at 51·8 knots.

Impressed by the Liberty engine, Mackay Edgar fitted two in his *Maple Leaf IV,* but issued no further challenges for the BI Trophy. Wood built *Miss America III* and *IV* in 1925 to face a French challenge from Henri Esdres, which did not materialise until 1926 when Wood built *Miss America V;* Esdres' *Excelsior-France* had to be towed around the course to start the engines and then the boat broke down after half a mile of the race. *Miss America V* won at 53 knots. There were no more challenges until 1928, Gar Wood's rumbustuous personality as well as his seemingly invincible boats proving too great a deterrent. But the Englishwoman Miss Barbara Carstairs, a Standard Oil heiress, who raced a notably successful 1½-litre class hydroplane *Newg,* a Sunbeam-engined Saunders boat, decided to challenge with two new boats, *Estelle I* and *Estelle II.* The first was a 7·9m (26ft) hydroplane designed by F. Hyde Beadle and built by Saunders. It shied away from the heavy approach of previous challengers and defenders, and was lightly constructed and fitted with one 652·8kW (875bhp) Napier Lion engine. *Estelle II* also had a Napier Lion but was only 6·4m (21ft) in length; the step followed the shallow-V contour of the hull and she proved to be a slightly better boat than the flat-V *Estelle I.* Wood took the opportunity of the challenge to build a new boat, *Miss America VI,* a 7·9m (26ft) hydroplane fitted with two 746kW (1000bhp) Packard engines. Fifteen days before the race the boat capsized and sank at high speed, breaking three ribs of Wood's sturdy co-driver and mechanic, Orlin Johnson. The engines were recovered from the bottom of the St Clair River and put into a new hull, *Miss America VII,* in time for the race on the Detroit River. On the second lap *Estelle II* ploughed into a wave and Carstairs and her mechanic were flung out. Wood, with Johnson, took the Trophy at 51·5 knots, and later set a world record over the measured mile at 80·6 knots.

In 1929 Barbara Carstairs challenged again with *Estelle IV* (*Estelle III* had been scrapped while under construction). She was a 10·7m (35ft) hydroplane, built of double diagonal mahogany in Carstairs' own boatyard on the Isle of Wight. She was fitted with three 746kW (1000bhp) Napier Lion 12-cylinder engines and was almost as fast as Wood, but on the third lap of the race she hit driftwood and retired. She was lapped by Wood's *Miss America VIII,* at 65·3 knots, and the following year *Estelle V* also lost to a new Wood boat.

England held the land and air speed records during these later years of American dominance afloat, and the land speed record holder, Sir Henry Segrave, became attracted to the idea of the water speed record. His car, the Sunbeam-powered 'Golden Arrow', was financed by Lord Wakefield's oil company, which was prepared to back an attempt on the BI Trophy. The man they went to was Hubert Scott-Paine (1891-1954), a pioneer aviator who had become head of Supermarine Aviation Works in 1920 and built the seaplane which won the Schneider Trophy in 1922. As a boy delivering paraffin from his father's ironmongery in Shoreham, Sussex, Scott-Paine had come to the notice of Noel Pemberton-Billing, whose yacht was kept in Shoreham harbour. He took on Scott-Paine and when Pemberton-Billing went into Parliament in 1916, Scott-Paine took over the Supermarine Aviation works, becoming managing director and owner four years later. He was known, partly from his ginger colouring and partly from his peppery temperament as 'Red-hot', and was the first person to employ R. T. Mitchell, taking on the aircraft designer straight from his apprenticeship.

In 1923 Scott-Paine was persuaded, apparently for £200,000, to leave Supermarine[3] and in 1924 he became a founder director of Imperial Airways. He was skilled in predicting, even shaping, public taste, and with his money from Supermarine he founded the British Power Boat Company in 1927 at Hythe on Southampton Water. He was intent on using flow production techniques to provide fast motor boats to meet the demand created in the mid and late 1920s by imported American boats, notably Chris Craft. Not least of Scott-Paine's attractions for Wakefield was his connection with the Napier company, whose Lion engines powered two Supermarine seaplanes to first and second places in the 1927 Schneider Trophy. The state of engine construction in Britain was lamentable; only Sunbeam, Rolls-Royce and Napier were producing engines of a reasonable power/weight ratio, the last two with government aid, with priority for aviation development. There were numerous firms making small petrol and diesel engines, such as Kelvin, Beardmore, Gardner, Meadows and Thornycroft, of up to 149·2kW (200bhp) in some cases, but the most powerful engines available were American: Scripps, Sterling, Cummins, Packard.

Segrave's boat, *Miss England*, was built by British Power Boat in 1928, to an advanced design by Fred Cooper. Cooper had been working up a reputation for hard-chine powerboats, mostly built by Saunders, but Scott-Paine periodically hired his talent. *Miss England* was an 8·4m (27ft 6in) single-step hydroplane fitted with a 693·8kW (930bhp) Napier Lion engine, adapted by Scott-Paine. Napier and Rolls-Royce were the only companies producing water-cooled aero engines, so the problems of conversion on the Lion were partly simplified and Scott-Paine provided a marine gearbox and skilful modifications to the exhaust manifold and the lubrication system. In a regatta held at Miami Beach in 1929 Segrave, who had just bettered his own land speed record, beat Wood's *Miss America VII* with *Miss England*, but this was chiefly because the scoring system mitigated against Wood's technical mishap in the first heat. Wood won the BI Trophy in 1929 (Segrave did not interfere with Barbara Carstairs' right to challenge) but at a meeting at the Lido in Venice in September 1929 *Miss England* conclusively beat *Miss America V* and *Miss America VII* at a speed of 79·8 knots. This was much closer to *Miss America VII's* world speed record of 80·6 knots and was largely due to Segrave fitting in to *Miss England* the 708·7kW (950bhp) Napier Lion from 'Golden Arrow'. The success of the lightweight and nimble *Miss England* led Segrave and Wakefield to consider a new boat of similar versatility but with more power to improve her chance of success in competition. *Miss England II* was entered for the 1930 BI Trophy with *Estelle IV* and *Estelle V*, but the sponsors' prime target was the world water speed record. Designed by Fred Cooper, she was a single-step hydroplane, of shallow-V form to the step, with a fairly flat run aft to the stern. Wakefield was able to procure two Rolls-Royce V12 engines, of the type used in the winning Supermarine S6 seaplane in the 1929 Schneider Trophy, and Cooper designed a gearbox to transmit 2611kW (3500bhp) from the engines, stepping up the speed to 12,000rpm on a single propeller. The boat was similar to *Miss England*, with her engines driving forward to a V-drive gearbox, the propeller strut combined with the rudder, and with fair lines to create a reasonably seaworthy hull. The 11m (36ft) boat was given its trials on Lake Windermere in June 1930 and on his record attempt Segrave achieved a world record of 85·7 knots, in the first boat to exceed 100mph. When *Miss England II* was on her third run she struck a log which tore a hole in her step which, because it was attached to and not an integral part of the hull, instantly

Miss England **was designed by Fred Cooper and built in 1928 for Sir Henry Segrave, at the helm, by the British Power Boat Company, whose proprietor, Hubert Scott-Paine, is on the foredeck of the racing boat. Scott-Paine's own considerable skills at handling and racing boats contributed substantially to his understanding of high-speed design.**

filled with water and capsized the boat. One of the mechanics and Segrave were killed, though he was told he had broken the record before he died. *Miss England II* was recovered from the lake and repaired, but not in time to compete in the BI Trophy. After this, Scott-Paine tried to secure rights to build a pair of the Rolls-Royce R-type engines for a BI Trophy contender he was considering. He was referred to Wakefield, who refused to let him have the engines from *Miss England II*, entrusting a new challenge to a man called Kaye Don. Like Segrave he had driven racing cars but he had less feel for handling boats; nevertheless, with *Miss England II* he raised the water speed record to 89·9 knots and then to 95·7 knots.

For the 1931 defence Wood had heavily supercharged the twin Packard engines of *Miss America IX* to 1044·4kW (1400bhp) each. He benefited considerably from Packard's continuing development of their V12 aviation engine, such as the US Navy's attempt at the Schneider Trophy which was made by a Curtiss powered by a Liberty engine. Wood paid Packard some $25,000 to assist in research in improving the marine version; most of the development work was carried out by Packard's chief engineer, Colonel J. G. Vincent. On the first heat on the Detroit River, six laps of a five-mile course, Don drove *Miss England II* to a win, with a new BI heat record of 78 knots and Wood was beaten for the first time in a Trophy heat, though he set a personal race record of 75·6 knots. For this competition *Miss England II* was fitted with a spiral propeller. In the second heat both Wood and Don were disqualified for jumping the starting gun, though neither knew it, and in attempting to round a buoy inside Wood, Don capsized. *Miss America IX* was flagged off the course, and the heat and the Trophy were won by George Wood in *Miss America VIII*. There was some denunciation of Wood after the second heat and he was charged with artifice in staging the disqualification, but this was unlikely, as he had adjusted the trim of *Miss America IX* and gained sufficient extra speed to beat Don on the second heat. Commenting, more or less in his defence, Barbara Carstairs said, 'I consider Mr Wood a hard opponent. He is naturally very keen to win, and does everything he can to do so. If Kaye Don was foolish enough to follow the American boat over the line, then it is his own fault.'[4] The affair rankled for a while but Wood got over it by raising the water speed record to 96·9 knots in Miami in 1932 and Wakefield commissioned Cooper to design another challenger. *Miss England III* was a more straightforward hydroplane, a 10·7m (35ft) boat built by Thornycroft with an integral step, two Rolls-Royce engines driving two propellers, and a transom stern instead of the deep-V counter stern of her predecessor. Don drove her to a new water speed record of 104·1 knots on Loch Lomond in July 1932. To meet this challenge Wood built *Miss America X,* a 11·6m (38ft) hydroplane with four 1193·6kW (1600bhp) Packard engines, driving in tandem two propellers. She was a faster boat than *Miss England II* but Don only failed because of a mechanical fault. Wakefield gave up challenging after this, unable to winkle any more engines out of Rolls-Royce, and in 1933 Wood defended the Trophy for the last time, against Scott-Paine, who pitted a 7·3m (24ft) aluminium alloy hydroplane against the burly *Miss America X*. Unable to buy an engine from Rolls-Royce, despite Air Ministry sanction, Scott-Paine resorted to a 1044·4kW (1400bhp) Napier Lion which he put in *Miss Britain III*, a futuristic-looking craft of 95·5 knots.[5] He could not match Wood's boat, which held the water speed record at 108·4 knots but, with far greater manoeuvrability and a slightly better power/weight ratio, Scott-Paine gave Wood one of his closest races; both boats ran faultlessly and Wood won at a record race speed of 76·5 knots.

The curious effect of this decade of intense competition was to stimulate naval boatbuilding in Italy, Germany and France, while British and American naval staffs felt that private industry was providing enough information about fast boats upon which they would be able to draw when necessary. The Admiralty fully realised the need to keep in touch with developments, though it was unable to provide incentive to industry. In 1921 the CMB base at Osea Island was transferred to Haslar, and for a couple of years was run in accordance with the proposals of the 1924 committee on CMBs, namely to maintain a nucleus of boats for training and to study tactics. At the committee's suggestion, tenders were requested in 1925 for a new CMB to keep abreast of technical progress, but were later cancelled. Had Thornycroft shown substantially more initiative in developing its basic design and offered a boat that would have become indispensable to the navy, the Admiralty might have been better impressed, but the firm was content to make random improvements and to better the payload of a vessel which had been designed for a specific function during the war. Its success as a boat far outweighed its performance as a weapon, though the two aspects became blurred. In 1926 the CMB operations at Haslar were closed down. Influential in naval thinking were two factors: the flying boat and the airborne torpedo. It was felt by naval staffs that flying boats, notably those of Short Brothers of Rochester, Kent, could amply fulfill all the patrol and reconnaissance work of the offensive motor boat, while such seaplanes as the Fairey Flycatcher and the Hawker Osprey saw considerable service as spotter planes and many were carried aboard cruisers. Both Royal Navy and Royal Air Force aircraft were able to carry torpedoes and, as far as daylight raids were concerned, there appeared little to choose between airborne and seaborne attacks on particular targets. For night attacks, though, the small torpedo boat, when intelligently used, remained a supreme but unexploited weapon.

In the early 1930s the British press carried arousing reports of high-speed torpedo boats being built abroad, notably in Italy, and to a lesser extent in France. In October 1931 the Admiralty was stirred to appoint the Coastal Motor Boat Committee to re-examine the situation and in 1933 it suggested that a new CMB be designed and built to keep abreast of developments, and in particular to stimulate high-speed engine design. Tenders were to be requested from suitable firms (J. S. White, Thornycroft, Yarrow, Saunders-Roe, Vosper, British Power Boat Company and Short Brothers) for a 16·8m (55ft) boat of not more than 14·2 tonnes (14 tons) displacement, with a speed of 45 knots, carrying two 45·7cm (18in) Mark M torpedoes (a model developed for the Fleet Air Arm) and two to three Lewis guns. These were exacting specifications. Aware of Admiralty thinking, Thornycroft offered a design in 1932 for an 18·3m (60ft) 38-knot boat, with two of their V12 engines, uprated to 335·7kW (450bhp). This proposal was not accepted because it was considered essential to be able to carry the new CMBs in ships' davits and Thornycroft's boat was too heavy. It was also realised that, apart from its seakeeping limitations, the intricacies of stepped hull construction were not suited to rapid production in the event of war.

The prevailing fashion, chiefly due to the spirited sales campaign of Scott-Paine, was for the more easily built hard-chine boat. However the Admiralty was chary of the upstart British Power Boat Company, despite being urged by the Air Ministry to try out one of its 11·4m (37ft 6in) seaplane tenders and a 4·9m (16ft) hard-chine dinghy. The Admiralty was in a quandary. On one hand it was fully aware of the need for keeping in touch with the progress of fast boats, yet was not prepared, or able, to give private industry the investment that would be necessary, maintaining that 'the CMB type of hull and machinery is kept alive to some extent under present conditions by the racing boats which are constructed from time to time'.[6] It was true that racing was a great stimulus to design, construction and machinery, but it offered little in the way of training of crews for active service or practice in firing torpedoes. On the other hand, there was the lacklustre service record of the CMB branch, the expense of maintaining a nucleus force, and the probability that aircraft could do the job just as well. Moreover, as a world power, Britain in the early 1930s had little to fear. It was believed, when considered at all, that Germany could be brought to its knees by economic measures. If there was any anxiety it was chiefly about Germany's growing air strength. The First World War might as well not have occurred for all the efforts Britain was making in the field of anti-submarine warfare, an area in which small craft could have made some contribution. The CMB committee reported in September 1933, but in October its recommendations were rejected and the tenders for a new CMB were not sought: there were no suitable engines; one boat would not have yielded sufficient information and experience to justify its initial cost; and if a boat or two were required for some special objective then there was sufficient boatbuilding experience in the country to produce a design to meet naval requirements.[7] The naval argument against rearming with small torpedo boats was unassailable. The Royal Navy did not need the boats and to have built, equipped, manned, and trained even a few flotillas would have been an extravagance costing hundreds of thousands of pounds.

The RAF required many small boats, however, to service its seaplanes and flying boats. Fast crash tenders at the RAF marine bases, such as Calshot, Felixstowe and Plymouth, were also considered desirable, though the flying boats had a fine record of safety. Thornycroft had built four launches for the RAF between 1927 and 1935, all fitted with Thornycroft V12s, but these ponderous engines were not considered economical to run,[8] nor was the firm

British Power Boat's 19.5m (64ft) air/sea rescue launch built in 1936 for the RAF. On a trial run from Grimsby to Southampton, 373 miles, this boat averaged 31 knots and was capable of 37 knots; she was faster than any boat then owned by the Royal Navy. Twenty-two of this type were built by 1940, when they were armed with Lewis guns, and in 1940-2 one hundred of a 19.2m (63ft) type were built. *Motor Boat and Yachting*

willing to supercharge them – the prevailing manner of boosting performance. In 1930 Scott-Paine offered for testing by the RAF Marine Branch an 11·3m (37ft) hard-chine launch of 23 knots powered by two 74·6kW (100bhp) Meadows petrol engines. Its hull lines owed much to Fred Cooper's influence, with a flared bow and hollow-V section forward sweeping aft to a flat transom. Cooper left British Power Boat in 1930, irked by the manner in which Scott-Paine had claimed so much of the responsibility for designing *Miss England,* and joined the Portsmouth engineering and boatbuilding firm of Vosper & Company. Scott-Paine had collared such a large proportion of Meadows production that other boatbuilders were handicapped by not being able to make use of the sturdy 4½-litre 6-cylinder engine from the Wolverhampton factory. He was in an unrivalled position to meet the Air Ministry order for a further eighteen fast launches. Scott-Paine was neither a naval architect nor an engineer, but he was able to sketch and explain his ideas, and correct any problems that occurred in practice or use with an unfailingly accurate empiric sense of design. This was a skill he had developed during his years in aviation. At Hythe at this time was T. E. Lawrence, sent there by the RAF (as Aircraftman Shaw) to cooperate in the development of Marine Branch launches. By 1934 the RAF Marine Branch felt the need for a larger boat, better able to put to sea than the 11·3m (37ft) launches, which were decidedly for inshore waters. In June 1935 the RAF had formulated its needs and requested tenders for a vessel of 11·3-21·3m (60-70ft) with a range of 500 miles and a speed of not less than 35 knots. A dozen firms were invited to compete, including Thornycroft, Vosper, J. S. White, Groves and Guttridge, James Taylor, British Power Boat, Camper and Nicholson and Brooke Marine of Lowestoft. White offered the cheapest boat but could only provide expensive Hispano-Suiza engines; Thornycroft submitted a stepless 21·3m (70ft) design but its installation of twin Thornycroft V12s gave a speed of only 30 knots. Other builders also experienced problems in finding suitable machinery. The Davey Paxman 559·5kW (750bhp) diesel was still under development and few were available. British Power Boat won the contract, and not just by virtue of being able to install Napier Lions. The 19·5m (64ft) hard-chine launch, flush-decked with a low wheelhouse amidships, had a range of 500 miles at 32 knots and 800 miles at cruising speed. It was fitted with three 373kW (500bhp) Napier Sea Lions and had a maximum speed of 40 knots in calm conditions. The hull was of mahogany; the side planking was screwed to ribands and the bottom planking was double diagonal mahogany through fastened with copper rivets. Scott-Paine's use of ribands (or battens) enabled his hard-chine craft to be produced in hundreds. On the early boats he had tried to use battens to fasten the bottom planking but customers quickly discovered that the system did not withstand the pounding duress of planing speeds, and the company was obliged to manufacture with sawn frames and stringers below chine level.

British Power Boat delivered twenty-two high-speed launches of the 19·5m (64ft) type to the RAF between 1936 and 1940. They were essentially for use in coastal waters but they were by no means unseaworthy, and in the Second World War air/sea rescue launch crews frequently ploughed through heavy weather in the North Sea and English Channel to reach ditched aircraft. Anxious to impress, Scott-Paine sailed the first ASRL on a 373-mile trial trip from Southampton to Grimsby, at an average speed of 31·4 knots. The unheralded difference between Scott-Paine's designs for the RAF launches, his motor torpedo boats which evolved alongside them, and the CMBs they were replacing, was that the new boats carried substantially larger crews and could keep the sea for forty-eight hours or more. This greatly altered the strategic rôle of fast fighting boats which, from their inception in the 1870s, had been equipped to execute one sortie and return to base. This principle had been reinforced by the hit-and-run CMBs, but now for the first time the mosquito fleet was wholly able to engage in action without any dependency on being towed or ferried by mother ships. This aspect was primarily a function of their larger size and the small amount of space taken up by the machinery. In the British Power Boat ASRLs there were quarters for a crew of eight in the forecastle and a cabin for four officers aft of the wheelhouse; there was sitting accommodation for twenty men in the hospital bay amidships.

British Power Boat was not alone in endeavouring to sell its wares to home and overseas governments. For many other companies there were full orders for hard-chine picket boats, admirals' barges, refuelling tenders, destroyers' boats, torpedo recovery vessels and target towing craft. For most of these boats, the question of speed was predominant, and the company that could offer the best performance was in a strong position to secure orders – hence British Power Boat's success in the early 1930s. The company had a near-monopoly of both Meadows and Napier output. With respect to the Napier Lion this was particularly galling to the Air Ministry, which had paid large sums for the development of the engine in the 1920s and was expected to buy afresh the British Power Boat marinised Sea Lion, substantially the same engine, in the 19·5m (64ft) ASRLs. Scott-Paine's agreement with the Napier company was the cause of some chagrin in Whitehall. He was regarded with a certain amount of unease by British naval staffs who were alienated by his wily and ostentatious marketing campaigns, yet his manifest talents resulted in the sale of the first hard-chine ship's boat to the Navy – a 4·9m (16ft) dinghy which had been bought in 1933 following its loan from the RAF Marine Branch; the outright purchase of the first planing admiral's barge in September 1934, which resulted from a direct approach by Scott-Paine to the Commander-in-Chief at Devonport; and the sale of the first fast admiral's barge to be carried aboard ship. The enthusiasm of the operating officers and crews easily overcame the disbelief of the authorities with regard to the performance of the boats. By the end of 1935 the Navy had bought forty fast boats from British Power Boat and taken delivery of hard-chine picket boats from J. S. White, Vosper and Thornycroft. Vosper's first craft for the Navy were 4·9m (16ft) hard-chine jolly boats, with clinker topsides and powered by a marinised 48·5kW (65bhp) Ford V8 petrol engine; they entered naval service in 1933 and were widely used.

Scott-Paine had outline plans for a fast torpedo boat as early as 1933 – when the British government had least intention of equipping the Navy – and he took the design to Estonia. After studying his proposals the Estonian naval staff then requested tenders from other boatyards, revealing the substance of Scott-Paine's ideas. He returned to England to develop his ideas further with a navy which had greater integrity.

Though interest in CMBs was attenuated in Britain from 1918 until the mid-1930s, research and development in Germany was intensive. The Treaty of Versailles 1919 restricted the number of German warships, forbade certain types and limited the displacement of future replacements. Forced into fresh thinking, German naval architects made excellent use of welding alloys and high-grade steels to save weight and build powerfully armed vessels of limited size. The pocket battleship *(Panzerschiff)* was the most dramatic result of the restrictions but the designs at the other end of the scale were equally impressive. The *Schnellbootswaffe*, the fast boat division of the German navy, was also a direct result of the arms limitations. Germany had been permitted a dozen small destroyers and twelve torpedo boats of not more than 203·2 tonnes (200 tons). Twelve destroyers were built between 1924 and 1929, but they were classified as sea-going torpedo boats (a class Germany had built in large numbers during the First World War) and by pro rata reduction, the permitted torpedo boats were only conceivable as motor boats. Accordingly, eight UZ boats (fast submarine-chasers) were built during 1919 and 1920; 30·8m (101ft) vessels of 14 knots powered by two 186·5kW (250bhp) petrol engines. In 1924 the navy co-opted seven of the fifteen LM boats still in existence to make a detailed study of future requirements. The boats were in the possession of Lürssen and in 1929 they were bought by the navy for further tactical evaluation and, more importantly, training. They were not armed, but to deter close inspection they were registered as submarine destroyers (for by this time it had been learnt how to interpret the rulings of Versailles, especially in boatyards).

Small torpedo-boat development was vigorously fostered by Captain S. Lohmann. He had supervised the clandestine technical and tactical growth, when to avert attention the boats were dubbed *Wachboote* (coastguard cutters), or speedboats (one such example was Lürssen's *Namenlas,* a 10·4m [34ft] hydroplane of 58 knots powered by two 194kW [260bhp] Maybach petrol engines), or were ostensibly motor yachts, such as *Narwal*, a multi-step 21m (69ft) cruiser with two 373kW (500bhp) Atlantic petrol engines, similar in appearance to one of Thornycroft's CMB-type yachts. Lürssen also built private yachts such as the steam turbine *Bremse* and *Brummer*, ordered by the Ministry of Finance, and the firm had kept up with developments by completing several wartime designs for other navies. A decisive design was the motor cruiser *Oheka II,* built for an American customer in 1926. She was a fast displacement boat of 34 knots, powered by three 410·3kW (550bhp) Maybach engines. The Reichsmarine was disappointed with hard-chine and stepped craft and saw that a vessel like *Oheka II* could be employed as a UZ or *Wachboot,* both permitted under treaty regulations. If built with torpedo scallops, to look like part of the streamlining, it would only need the addition of standard torpedo-tubes to serve as a light torpedo boat when the treaty was repudiated. Lohmann's direct involvement ended in 1926 when he was embroiled in a political scandal, but the substance of his work was taken up and continued by the Lürssen yard and the naval architect Alfred Bunje, who were jointly responsible for most S-boat production until 1945.

In November 1928 the navy asked Lürssen for the plans of *Oheka II,* and it returned them, four months later, with certain amendments. Then in November 1929 the navy ordered the modified boat from Lürssen, *UZ* (S) *16*, later renumbered *W 1* and then *S 1*, and the boat was built and commissioned by August 1930. She was 27m (88ft 6in) long, with a beam of 4·2m (13ft 9in) and a draught of 1·1m (3ft 6in). The hull was built in double diagonal mahogany on alloy frames, and three 671·4kW (900bhp) Daimler-Benz V12 petrol engines were fitted, on steel engine-bearers. She was a 34-knot boat of 39·6 tonnes (39 tons) displacement and could keep the sea in sea states 5-6. Two 50cm (19·7in) torpedo tubes were located on the foredeck, to fire torpedoes over the bow.

Meanwhile three experimental motor torpedo boats had been built between 1925 and 1928, a single-step and a multi-step hydroplane, both fitted with stern torpedo tubes, and a riot boat of displacement form built by Lürssen. The 16·8m (55ft) single-step boat, produced by the sailing yacht builders Abeking and Rasmussen, was capable of 40 knots with two 320·8kW (430bhp) petrol engines.

By 1932 the 1st S-boat Flotilla (*S 1* to *S 5*) had been formed and under *Kapitänleutnant* Erich Bey had notched up scores of hours' operating experience in all conditions. The boats' performance was distilled into an improved design, *S6,* which formed the basis of all subsequent classes. The most significant feature was her engines: three 984·7kW (1320bhp) 7-cylinder MAN diesels. The progress of diesel engineering had been rapid since 1913, when Dr Rudolph Diesel had allowed to lapse the patent specification which ruled that all diesels operate at a constant pressure. Diesel ignition is by high-pressure combustion, necessitating a substantially heavier engine than a comparably powered petrol unit, and some of the earliest diesels had a power/weight ratio of up to 204·3kg (450lb) per brake horse power. Many countries built large and powerful diesels after the First World War, for land and marine use, of 15-500rpm but only Germany made any progress with lightweight high-speed diesels of 500-1200rpm. The advantages of the marine diesel were considerable: the fuel was much safer to handle than petrol and because it was directly injected into the engine the drawbacks of carburettors were avoided; the fuel was ignited by compression alone, and so by-passed the necessity for electrical ignition. These two factors made the diesel an infinitely more reliable engine than the petrol-driven one, though diesels were much noisier in operation, required special starting apparatus, and it was to be many years before their power/weight ratios were brought to a level comparable with petrol engines.

The MAN diesels in *S 6* gave a top speed of 35 knots with a cruising range of 760 miles at 22 knots and the same engines were fitted in three more S-boats by 1935. The next four boats had Daimler-Benz V16 diesels, of the same power but greater reliability, and though MAN produced a 1492kW (2000bhp) 11-cylinder engine for *S 14* to *S 17* in 1936 it was structurally unsound and thereafter only Daimler-Benz V20 diesels were installed in S-boats.

One of the first *Schnellboote, S 4,* built in 1932 and capable of 34 knots. She was one of the five S-boats fitted with triple 820.6kW (1100bhp) Daimler-Benz petrol engines, and which were discarded before the Second World War. In addition to the 20mm AA gun she carried a small machine-gun on her foredeck. *P. A. Vicary*

The diesel-powered boats were armed with two 53·3cm (21in) torpedo tubes forward and a 20mm cannon aft. From *S 14* onwards the boats were increased in size to an overall length of 34·6m × 5·1m × 1·7m (114ft × 16ft 9in × 5ft 6in), displacement was 94 tonnes (92·5 tons), and most carried two 20mm cannon. Within ten years, the German navy had created a consummate torpedo boat: an offensive sea-going vessel capable of sustained high speeds in the Baltic, North Sea and English Channel. The lines of the S-boats were, and remain, a near-optimum design for the purpose. All S-boats had displacement hulls, with quite a hard turn of bilge and flat floor sections from amidships to the transom; the forward sections were softer, with moderate flare at the bow and a chine above the waterline ran from the stem to amidships to deflect spray. As the performance of the boats increased during the 1930s the mahogany planking was replaced by double or even triple skinned teak, on alloy frames, to add to the strength of the hull. The boats with triple 1492kW (2000bhp) engines were capable of 39 knots. The S-boats were notably level-riding craft, and a factor contributing to this was discovered during the trials of *S 2* when, at full speed, the helm was put hard over; instead of changing course the boat continued ahead at an increased speed of 1½ knots. To take advantage of this, small auxiliary rudders were fitted adjacent to the propellers and at high speeds they were turned outwards by 10-15 degrees, with the same paradoxical increase in speed. The effect of the rudders at high speeds was to reduce the tendency of over-powered hulls to squat deeply into the water at the stern, and

This S-boat of 1935 was fitted with triple 1492kW (2000bhp) MAN diesels, producing a speed of 40 knots, but the engines were unsuitable and after 1939 (*S 26* onwards) all S-boats had Daimler-Benz engines.

so to reduce the wavemaking resistance. It led Lürssen and Bunje to experiment further with trim control and proceed to fit horizontal as well as vertical planes, and some S-boats had wedges built on to their trailing edges: forerunners of the hydraulically controlled transom flaps fitted to many modern fast patrol boats.

The S-boat was an expensive vessel, and strategically it was rarely regarded as a disposable hit-and-run boat, such as the CMB, but was seen essentially as a small destroyer. They were administered by *Konteradmiral* (Rear-Admiral) Butow, *Führer der Torpedoboote* – an indication of the standing in which Germany held S-boats. Their rôle was to attack warships in coastal waters, though merchantmen were not excluded from their brief and during the Second World War became S-boats' prime targets. The boats were painted off-white for night-time camouflage and for quiet-running were fitted with a heavily silenced 74·6kW (100bhp) Maybach engine. Many exercises during the 1930s were devoted to silent stalking of their targets.

British naval staffs were woefully indifferent to the research, construction and training which was taking place in Germany during the 1920s and 1930s. As a cover-up operation it lacked conviction, and in the 1931 edition of *Jane's Fighting Ships* Lürssen went so far as to advertise, with a photograph of *S 1* captioned as a 35-knot 'displacement speed boat'. But there was no attempt to acquire an S-boat for evaluation. Indeed, during the war the Royal Navy was most anxious to capture an S-boat, mindless of the report made by a British naval attaché who went to sea in one in 1937, and of other technical information which had been published about them in Britain. This neglect of rival naval progress was due to the continuing belief that the torpedo boat was the weapon of the weaker power, and that Germany was only following form in building up flotillas of little ships.

6 EMERGENT NOTIONS

It is to the everlasting credit of Scott-Paine that in October 1934 he was able to persuade the Admiralty to place an order for two British Power Boat motor torpedo boats. He had the sympathy of the Third Sea Lord, Admiral Sir Reginald Henderson, who pressed for the boats' adoption in the Navy, and of the Assistant Chief of Naval Staff, Admiral Kennedy-Purvis, and his successor, Captain Woodhouse. The two boats were regarded as experimental and were to serve in the Mediterranean but four more were added to the order when the Italian invasion of Abyssinia began in October 1935. There were few people who doubted the value of coastal motor boats (see the findings of the 1933 CMB committee[1]) but it would not be too critical of the Admiralty to say that it was reactionary as far as coastal motor boats were concerned. To most naval minds the idea of a small fast motor boat putting to sea with torpedoes still represented a substantial alteration to their world – a way of life dominated by capital ships, submarines and destroyers. Moreover, such boats had an indifferent service record and many were unseaworthy and unreliable. Even the old arguments in their favour were no longer valid: they were not cheap to build (the Scott-Paine boats cost £23,000 each compared with £14,000 for a new CMB); with petrol engines instead of steam engines they were far from stealthy; and in 1935 there was no boatyard able to produce rapidly the swarms of, by then, quite complicated torpedo boats. Most telling of all was that they were no longer the pace-setters of the modern navy. Scott-Paine's torpedo boats were 3 knots slower than the 36-knot destroyers of the mid-1930s. They were also slower than the 40-knot CMBs of the First World War, despite having nearly three times as much power – but they carried a greater payload, had a greater range, and could cope with rougher conditions than the CMBs.

All six boats were completed in 1936 and commissioned in June 1936. They were substantially the same hulls as British Power Boat's 19·5m (64ft) RAF boats, though they were only 18·3m (60ft) in length and had aluminium decks. There were quarters for nine men, including two officers. The fuel tanks for 2270 litres (500 gallons) were beneath the central deckhouse and the engine room occupied the last third of the boat's length. This was because of the unique arrangement for firing the torpedoes, which were carried on overhead rails in the engineroom. Before reaching torpedo range (800m [800 yards] was the ideal) the boat had to slow down, lift the lattice-work girders from their stowed position and hinge them over the transom, forming a continuation of the engine-room rails. The chocks were removed and the boat suddenly accelerated, so that the torpedoes ran aft and overboard. Their motors had been trip-started and the launching MTB swerved to one side to give the torpedoes a clear run to the target. It was an inexpensive and lightweight system, and anything of greater complexity would have meant that the boat had to be bigger, and so more powerful. It had been worked out in collaboration with HMS *Vernon* (the Navy's torpedo school) who contributed about 25 per cent of the development work (a matter of some contention when Scott-Paine patented the method). However, it was not a satisfactory arrangement. From the removal of the stops until the torpedo was in the water took at least six seconds and the special rollers welded to the weapon to guide it in the cradle reduced its performance by 30 per cent. The most serious disadvantages were that the torpedoes could not be fired when the boat was stationary or moving at slow speeds – the circumstances most conducive to a successful attack – and that to be fired at all the boat had to be travelling towards its target at such a speed that it could accelerate and so eject the torpedo: onerous conditions to impose upon a boat's commander. Scott-Paine did have a subsequent idea for launching the torpedoes nose-first by elastic cord from a pocket on either side of the hull, close to the waterline.[2] The Admiralty was not prepared to pay Scott-Paine a royalty, pointing out that this method was little more than an amalgam of tried techniques.

The first six boats had three 373kW (500bhp) Napier Lions, giving a maximum speed of 33 knots and a top continuous speed of about 29 knots. In addition to the two 45·7cm (18in) Mark VIII torpedoes (no re-loads were carried in British boats during the Second World War though it was standard practice aboard S-boats), two clusters of four 0·303in machine guns were mounted on slip rings; one on the foredeck and one at the stern. The boats were

Boats of the 1st MTB flotilla on the Thames in May 1937, the first vessels of their type in the Royal Navy. The hinged covers on the transoms (right) are the ports through which the boats' torpedoes were launched when the girders were rigged over the stern (below and right).

extremely light, displacing 22·4 tonnes (22 tons) in service trim, though the weight-saving techniques led to trouble and were subsequently amended. The bottom planking was double diagonal mahogany and the side planking was single thickness, screw fastened to seam battens. The sawn frames were 0·5m (1ft 6in) apart, a reasonable spacing for light sea conditions, but the boats were not handled lightly and when driven hard the frames worked and fractured and the screw fastenings loosened. Subsequently British Power Boat had to build their torpedo boats with clenched double diagonal side planking and frames at 0·3m (1ft) intervals. A second batch of six boats was authorised at the board meeting of the 1936 Construction Programme Committee in January, and the contract was given to British Power Boat in May 1936, for delivery in six months.

It was at the January 1936 meeting that the boats, which had invariably been referred to as coastal motor boats, were reclassified as motor torpedo boats. The 1st MTB Flotilla of six boats worked up in the spring of 1937. In June 1937 they made the passage across Biscay to Gibraltar and Malta on their own keels, and ahead of schedule. This won considerable acclaim, though it was a perfectly straightforward exercise and they were in company with HMS *Vulcan,* a trawler converted into a depot ship. (In fact their return journey through the French canals was far more arduous than the open sea route.) During service at Malta the boats' shortcomings in the way of armament, speed and endurance became apparent. The 2nd MTB Flotilla was completed in 1937 and shipped to Hong Kong and an order for six more 18·3m (60ft) MTBs was again placed with British Power Boat, bringing the total to eighteen boats. The 3rd flotilla was completed in 1939, by which time the cost of the boats had risen to £38,000 each.

Once committed to an MTB programme, the Admiralty did persevere in trying to find the right formula; the failings of the 18·3m (60ft) MTBs were held to be due to the deficiencies in the boats rather than Admiralty policy. With the odour of war becoming slightly more pungent, the Admiralty decided in October 1936 to fix four areas of fast boat development: anti-submarine boats; MTBs for ad hoc operations; a type of MTB for local defence and a seagoing MTB able to stay with battle fleets in narrow waters. Arrangements for the first three were in hand but the need for a seagoing MTB was urgent because it was widely believed that otherwise a destroyer would have to be attached to each MTB flotilla.

As had been found with CMBs it was folly to use fast boats for anti-submarine duties. The vessel required for this sort of work was a sturdy displacement boat, well able to keep the sea in the North Sea and English Channel. This work was valiantly performed during the Second World War by trawlers, but five motor anti-submarine boats were included in the 1938 construction programme, in place of one patrol vessel. The MASBs, known as Masbies, were 18·3m (60ft) British Power Boat designs, carrying depth charges, with two Napier Lions, giving a speed of 25 knots. They were fitted with asdic and hydrophones and because of the heavily screened electrical circuits the transducer, mounted 6·1m (20ft) from the bow, was found to pick up underwater signals quite satisfactorily, though it was trained only by steering the boat. This technical success encouraged further building of MASBs, of 19·2m (63ft) in length, though the first five were stationed at Portland in 1939 defending the harbour. In November 1938 British Power Boat produced a 21·3m (70ft) MTB and fifteen 19·2m (63ft) MASBs were ordered, structurally akin to the larger hull and powered by two Napier Lions. The MASBs were thought to have an important rôle repelling U-boat activity in the Channel, but it was uneconomic to operate hard-chine boats at sub-planing speeds because the engines use almost as much fuel pushing a bulky hull through the water at 8-12 knots for long periods as in planing over the water at 22 knots. Of greater merit was Thornycroft's design for motor minesweepers. Two were sanctioned by the Admiralty and built by Thornycroft in 1937. They were 22·9m (75ft) craft powered by triple Thornycroft 354·4kW (475bhp) V12s and were capable of 15 knots. When sweeping, their speed was reduced to 10 knots. In 1939 the boats were sold to Turkey and the Admiralty started a construction programme of trawler-type 167·6-tonne (165-ton) motor minesweepers of 11 knots.

With the 18·3m (60ft) MTBs consigned to defensive operations, the Royal Navy was still in great need of a seagoing MTB and the competition among companies to design and build one was intense. The centre of English yacht and boatbuilding, for so long on the Thames, had gradually shifted to the Solent. Many great yards – Camper and Nicholson, Thornycroft, Saunders-Roe, Groves and Guttridge, J. S. White, Vosper, British Power Boat, and dozens of smaller builders – were scattered about Southampton, the Isle of Wight and Portsmouth (though most of Thornycroft's fast boats continued to be built at the Hampton yard on the Thames). It was an area rich in designers, shipwrights and engineers, most of whom had a shrewd idea of what their rivals were turning out in the way of work. The Royal Navy's interest in MTBs was increasingly represented by HMS *Vernon* at Gosport, in Portsmouth Harbour, and companies had the opportunity of public, or sometimes private, viewing of what was afoot. The competition to secure service orders was frequently keen and often virulent.

The question of a multi-rôle MTB to succeed the CMB was considered to

Completed in 1939, White's 42-knot experimental hydrofoil, *MTB 101*, was powered by three Isotta Fraschini engines and armed with two 53.3cm (21in) torpedoes and two 20mm cannon. Here she is passing a coaster while undergoing speed trials off Cowes. *Beken*

be in hand by J. S. White of Cowes, the oldest shipyard on the Admiralty's list, which had an experimental MTB on the stocks in 1937. J. S. White, like Thornycroft, had been busily employed between the world wars in building destroyers. Prior to 1937 White had built a small single-screw hydrofoil and, impressed by its possibilities, boldly decided to construct a hydrofoil MTB. The boat was completed in 1939, a 20·5m × 4·4m (67ft 3in × 14ft 6in) hull which drew 0·9m (3ft), excluding the foils. She was fitted with three 746kW (1000bhp) Isotta Fraschini petrol engines but was not notably fast; she had a continuous maximum speed of 36 knots and her best speed was 42 knots. The Admiralty expected much from this boat, *MTB 101,* but it was an expensive venture (she cost the Royal Navy £51,000) and her performance was substantially due to lightweight construction. Hydrofoils offered the operational advantage of good acceleration from rest or low speeds, but in the bustle of docking the protection of the foils became a constant anxiety. (Damage to stern gear, frequently through careless handling, contributed heavily to the amount of time the Royal Navy's small boats spent out of the water for repairs.) Unfortunately *MTB 101* was lost in 1942 when one of the foil struts collapsed.

In August 1936 the Admiralty suggested to British Power Boat that an outline design might be prepared for a boat capable of carrying two 53·3cm (21in) torpedoes and with at least the same performance as the 18·3m (60ft) boat. The question of torpedo armament was still under review. Since 1935 *Vernon* had been unhappy about the stern firing arrangement in the 18·3m (60ft) boats. The system had been acceptable at the time as offering a lightweight and cheap installation, but the 45·7cm (18in) Mark VIII torpedo had too small a warhead. The Royal Navy wanted the 45·7cm (18in) Mark IX, with a 199·8kg (440lb) warhead and preferably the 53·3cm (21in) Mark VIII which carried 340·5kg (750lb) of explosives. These larger torpedoes could only be fired from tubes, which added considerably to the weight of the boat. In 1937 Vosper built a 20·7m (68ft) tender for *Vernon* for use in testing torpedo launching and firing equipment. She was a 25-knot boat, powered by two 596·8kW (800bhp) Lorraine petrol engines, of French manufacture. Called *Bloodhound,* she could carry three torpedoes, and a tube was mounted on a turntable amidships which could be trained to port or starboard.

Thornycroft continued constructing its CMBs for overseas governments during the coming out of the MTBs but the company was unable to arouse Admiralty interest. The performance of the CMBs had improved over the years. Thornycroft had increased the output of their V12 engine from 279·8kW (375bhp) to 484·9kW (650bhp) by 1938, and the boats were able to achieve speeds of 45 knots with a full payload of two 45·7cm (18in) torpedoes and light armament. In 1939 Thornycroft was building 16·8m (55ft) CMBs for China, Finland and the Philippines but, in the enlarged rôle of the MTB, there was no place for the CMB. It could not provide a stable gun platform and it was a difficult type to handle in following or quartering seas. In 1940 Thornycroft produced four experimental stepped boats *(MTBs 104-107)* built in double diagonal mahogany and 12·2 to 15·2m (40 to 50ft) long, but they did not lead to the CMB orders for which the company had hoped. One of the boats *(MTB 105)* had a Rolls-Royce Merlin engine and a 45·7cm (18in) swivelling torpedo tube on the aft deck.

Perhaps the most expedient project in the pre-war years was the design by F. Gordon Pratt of Cox and King for a 33·5m (110ft) vessel called *Tarret.* She was built in 1939 by Swan, Hunter and Wigham Richardson of Wallsend-on-Tyne, Newcastle, of all-welded steel construction. *Tarret* displaced 116·8 tonnes (115 tons) and was powered by two Davey Paxman 746kW (1000bhp) diesels. She had a maximum speed of 30 knots, without armament, but could maintain this speed in rough weather. The hull shape was a development of

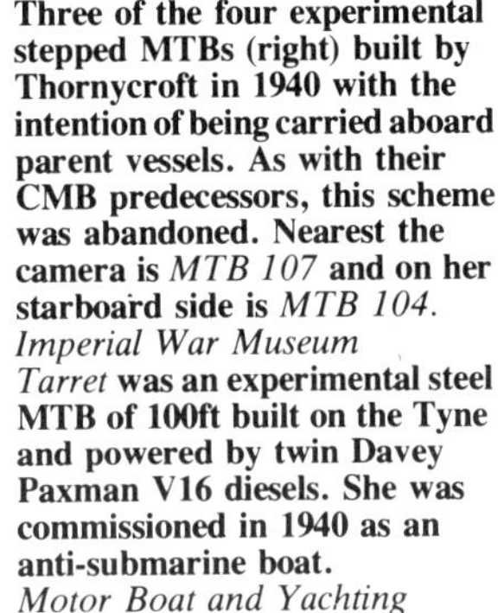

Three of the four experimental stepped MTBs (right) built by Thornycroft in 1940 with the intention of being carried aboard parent vessels. As with their CMB predecessors, this scheme was abandoned. Nearest the camera is *MTB 107* and on her starboard side is *MTB 104.*
Imperial War Museum

***Tarret* was an experimental steel MTB of 100ft built on the Tyne and powered by twin Davey Paxman V16 diesels. She was commissioned in 1940 as an anti-submarine boat.**
Motor Boat and Yachting

Pratt's designs of 1912, a quasi-planing form with severely concave sections inboard of the chine, reversing at the turn of a second chine to convex sections.[3] The outer chine forecast a future pattern by rising from the waterline at midships to the top of the stem. Higher speeds than *Tarret*'s were thought necessary for a seagoing MTB and it was considered that lightweight diesels of a suitable power/weight ratio could not be produced in time. *Tarret* was commissioned as an MASB in 1940 and used later for training.

British Power Boat completed its larger prototype MTB in November 1938. It was a 21·3m (70ft) boat of good performance and shapely appearance. The firm claimed she could mount four 45·7cm (18in) torpedo tubes or two 53·3m (21in) tubes and subsequent British versions carried just two of the larger tubes. She was powered by three Rolls-Royce 746kW (1000bhp) engines, the 27-litre V12 Merlin, a superb engine designed and produced entirely as a private venture between 1932 and 1936. Aviation requirements took precedence over the marine use of the Merlin from 1939 onwards, but Scott-Paine acquired a supply of the engines and the Rolls-Royce marinisation was carried out in cooperation with British Power Boat while the gearboxes were made by the Meadows factory of which Scott-Paine had become a director. This boat was offered to overseas navies and orders were placed by France, the Netherlands, and Sweden (for a 19·2m [63ft] version). However, after conducting trials the Royal Navy turned down the British Power Boat 21·3m (70ft) MTB. It was undoubtedly a great improvement on the 18·3m (60ft) type in matters of seaworthiness, equipment and habitability and it had a speed of 44 knots on trial, 40 knots maximum continuous with a trial load. The construction was double diagonal mahogany and in the forward sections the frames were spaced at 25·4cm (10in). The frames were bracketed at either side with plywood gussets at the gunwale, chine, and keel, and in heavy seas these were found to work loose. Teething problems of this nature were not unexpected, nor were they insurmountable, but Scott-Paine's personal fortunes were at something of a low ebb in 1938. Then in March 1938 his firm was accused, during the Parliamentary debate on the Navy supply estimates, of underhand dealings with the Admiralty.[4]

The charges drew attention to British Power Boat's apparent monopoly of MTB contracts and that the firm was charging new prices for second-hand Napier engines which had been bought at knock-down prices. The large order given by Imperial Airways for crash tenders was also cited and notice drawn to the fact that Scott-Paine was a director of Imperial Airways, a matter of concern in so far as the airline was a government-subsidised firm. In his reply the following month the First Lord of the Admiralty, Mr Duff Cooper, pointed out that it was only fair

> 'that the original designer of this type of boat should have received initial orders proportionate to this enterprise and to the success of the design, and it can be categorically stated that it is the intention of the Admiralty to invite competitive tenders for motor torpedo boats the moment they feel they are in a position to do so.'[5]

Just after this, in May, an order for nine boats was placed with British Power Boat and at the same time the Admiralty confirmed its intention of proceeding with Vosper's MTB design. British Power Boat had of course bought cheaply some used Napier Lions from the Air Ministry, to use in designing a suitable marine adaptation for the MTBs. Three of these engines had been sent to HMS *Vernon* for testing, when water entered the cylinders of one of them, necessitating reboring. It was this which had led the MP for Nuneaton, Lieutenant-Commander Fletcher, to berate the government. The following year the captious Fletcher raised the matter again, stating that after his allegations the Admiralty had not held a proper inquiry into how inferior engines could pass the normally strictly conducted acceptance procedures but had simply 'conducted an inquiry into their own conduct.'[6] After converting six second-hand Lions, British Power Boat was able to give Napier its specification for new engines; and it was the new engines which were installed in the MTBs. British Power Boat was exonerated but disparagement continued for some time.

Scott-Paine suffered the pioneer's fate of being superseded by competitors. His achievement was in persuading the Admiralty to take his boats when the Royal Navy did not know what to do with them, and the Admiralty's deferment of policy on small craft is wholly understandable in view of the activity among other boatyards to produce superior MTBs. In the hope that one of the firms with suitable experience might have come up with a winner, the Admiralty was not pressed to finalise requirements for a suitable MTB. Fortunately, a prototype which embodied many of the latest requirements was produced by Vosper in 1937. Vosper & Company had been reinvigorated in 1931 when Commander Peter Du Cane was appointed managing director. As a midshipman and naval engineer he had acquired considerable experience of the Royal Navy's small boats and their deficiencies, and was able to consolidate Vosper's reputation for small boats with a variety of fast jolly boats, mostly of the 4·9m (16ft) type. These were powered by a marine conversion of the 4-cylinder Ford petrol engine of 29·8kW (40bhp). Vosper also invested in marinising the Ford 48·5kW (65bhp) V8, a reliable and versatile unit which was used in numerous craft.

In 1934 Du Cane became responsible for the design of Vosper's boats and in 1935 interested the Admiralty in large numbers of a 7·6m (25ft) picket boat, and a year later in 10·7m (35ft) and 13·7m (45ft) versions, all powered by the Ford V8. Du Cane was a more discreet publicist than Scott-Paine and Vosper did not overlook the value of private customers; among their varied yacht work was a 32m (105ft) diesel-powered motor yacht built in 1935. It is a noteworthy tribute to Du Cane's accomplishment as a naval architect that, as the Admiralty was well aware, 'with the development of high speed racing motor boats since the war there is, it is believed, the knowledge in this country to build the boats [MTBs] required'.[7] Vosper had no experience of racing boats, apart from a pair of hydroplanes designed by Fred Cooper, who had also designed a few larger fast hard-chine boats, including *Advance,* a 12·2m (40ft) picket boat, well liked by crews on HMS *Hood* to whom she was loaned in 1932. But she was too radical and expensive an improvement on existing boats for the Admiralty to purchase. Cooper was with Vosper from 1930 to 1933. Du Cane's mastery of his work was sharpened by his years in the Auxiliary Air Force; observation of the critical nuances of trim and the dynamics of air flow were not wasted in designing fast boats.

By the mid-1930s Vosper was well poised to exploit the necessity for an MTB superior to the utilitarian but inadequate British Power Boat 18·3m (60ft) type. The Admiralty wanted a 40-knot boat which would carry two 53·3cm (21in) torpedoes and maintain its maximum operating speed in a fresh breeze. The key to the equation was in providing sufficient power; with that accomplished, the weight of the armament and design of a suitable hull became soluble problems. Vosper's answer was the Isotta Fraschini engine, an excellent unit, built in Milan, which had three banks of six cylinders. The engine had a good power/weight ratio and produced 857·9kW (1150bhp) at 1800rpm. It had undergone steady development in MAS-boats since 1929 and was renowned for its reliability. Vosper's private venture MTB was launched in

May 1937, powered by three Isotta Fraschini engines. She was 20·7m (68ft) in length, 4·5m (14ft 9in) in beam with a draught of 1m (3ft 4in). Her displacement was 32·5 tonnes (32 tons) at which she had a maximum speed of 43·7 knots and a maximum continuous speed of 35·5 knots. The hull construction was triple-skinned mahogany bottom planking and double-skinned side and deck planking, screw fastened to sawn mahogany frames. (Today mahogany is a precious timber but before the war it was used extensively in boatbuilding.)

The engine-room layout was the archetypal configuration of port and starboard units driving the wing propeller shafts, with a V-drive between them to reverse the drive shaft from the third engine mounted further aft. On either side of this central engine was a Ford V8, which could engage the wing shafts when silent – or at least much quieter – running was necessary. These auxiliaries were also used for low speed manoeuvring. Vosper was conscious of the disadvantages of the torpedo launching arrangements on the 18·3m (60ft) MTBs, and a means of firing a 53·3cm (21in) torpedo through a port in the bow was devised. A forward-firing torpedo had not been seen on British boats since the days of *Lightning* and it heralded a minor breakthrough in MTB tactics. One re-load torpedo was carried, and access to the downward-aiming torpedo tube was on a track aft of the deckhouse. When in rough seas, the forward torpedo was liable to be struck on its exit by the pitching motion of the bow, spoiling the torpedo's run. The torpedo armament was developed in alliance with *Vernon,* whose staff were also concerned about the delay between firing the loaded and the re-load torpedos, which it was thought would be a handicap when the boat was in action. Accordingly the loading rack was re-fashioned to launch the second torpedo tail-first over the stern. Thus the boat was fit to meet the Admiralty requirement for an MTB carrying two 53·3cm (21in) torpedoes.

The single feature of the boat most attractive to the Royal Navy was its hull shape. It had better sea-keeping qualities than the 18·3m (60ft) MTBs and held out hopes that a sea-going MTB might become a reality. The form of the British Power Boat MTB was little changed from the lines of the firm's boats designed by Fred Cooper: shallow deadrise; low chine line; concave forebody sections twisting to flat or slightly domed stern sections. The hollow bow and low chine were the cause of the notorious pounding but, without powerful engines (or building stepped hulls), they were efficient means of attaining good planing speeds. With the abundantly powerful Isotta Fraschini engines, Vosper could afford to compromise the high-speed factors and build in superior ride charac-

Vosper's private venture 20.7m (68ft) MTB on speed trials in Southampton Water in May 1937. She was capable of 40 knots and later demonstrated her seaworthiness in a Royal Navy trial off the Isle of Wight during a Force 7 gale. *Vosper*

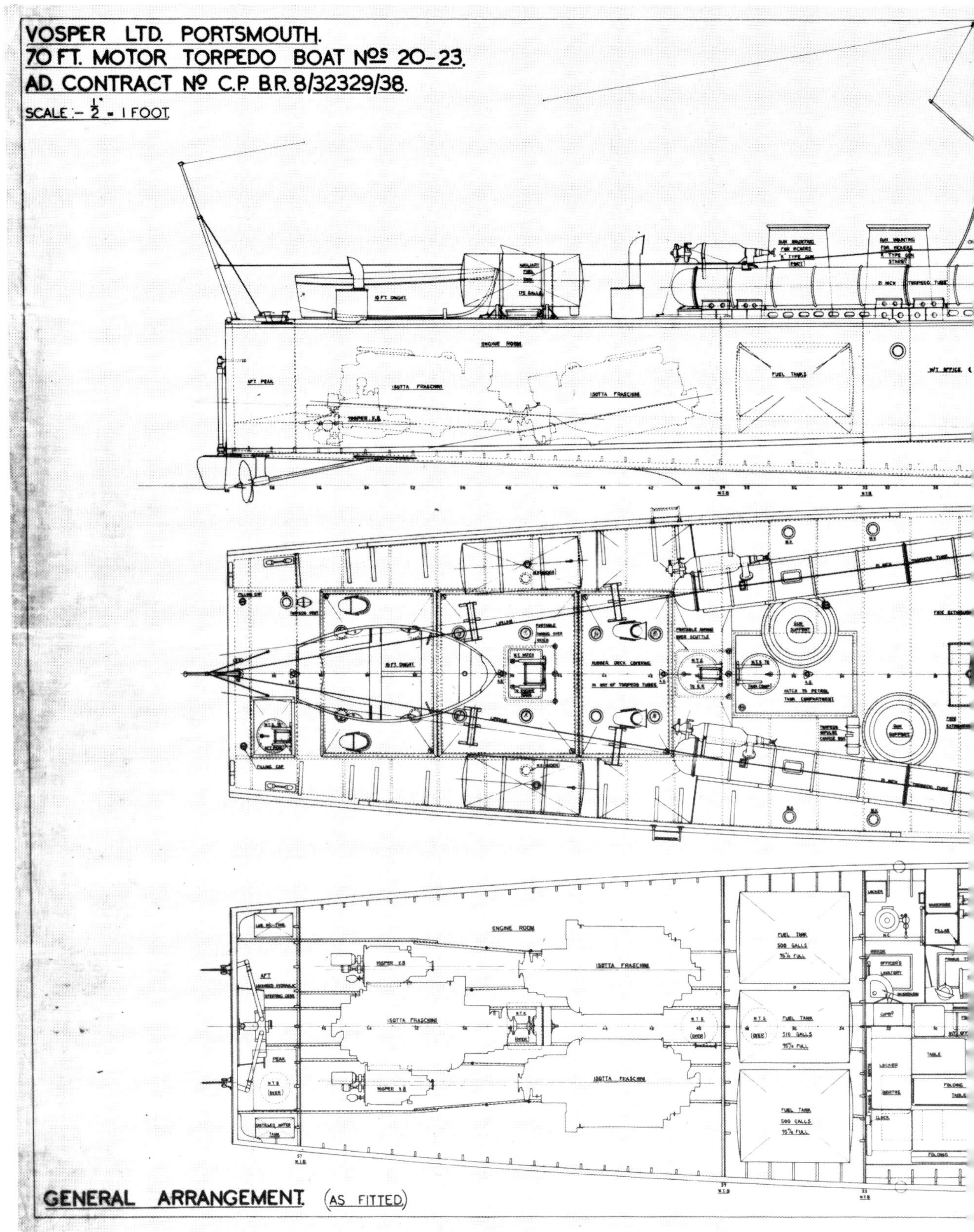
VOSPER LTD. PORTSMOUTH.
70 FT. MOTOR TORPEDO BOAT Nos 20-23.
AD. CONTRACT No C.P. B.R. 8/32329/38.
SCALE :- 1/2" = 1 FOOT.
ENGINE ROOM
AFT PEAK
ISOTTA FRASCHINI
FUEL TANKS
FUEL TANK 500 GALLS
FUEL TANK 510 GALLS
GENERAL ARRANGEMENT. (AS FITTED)

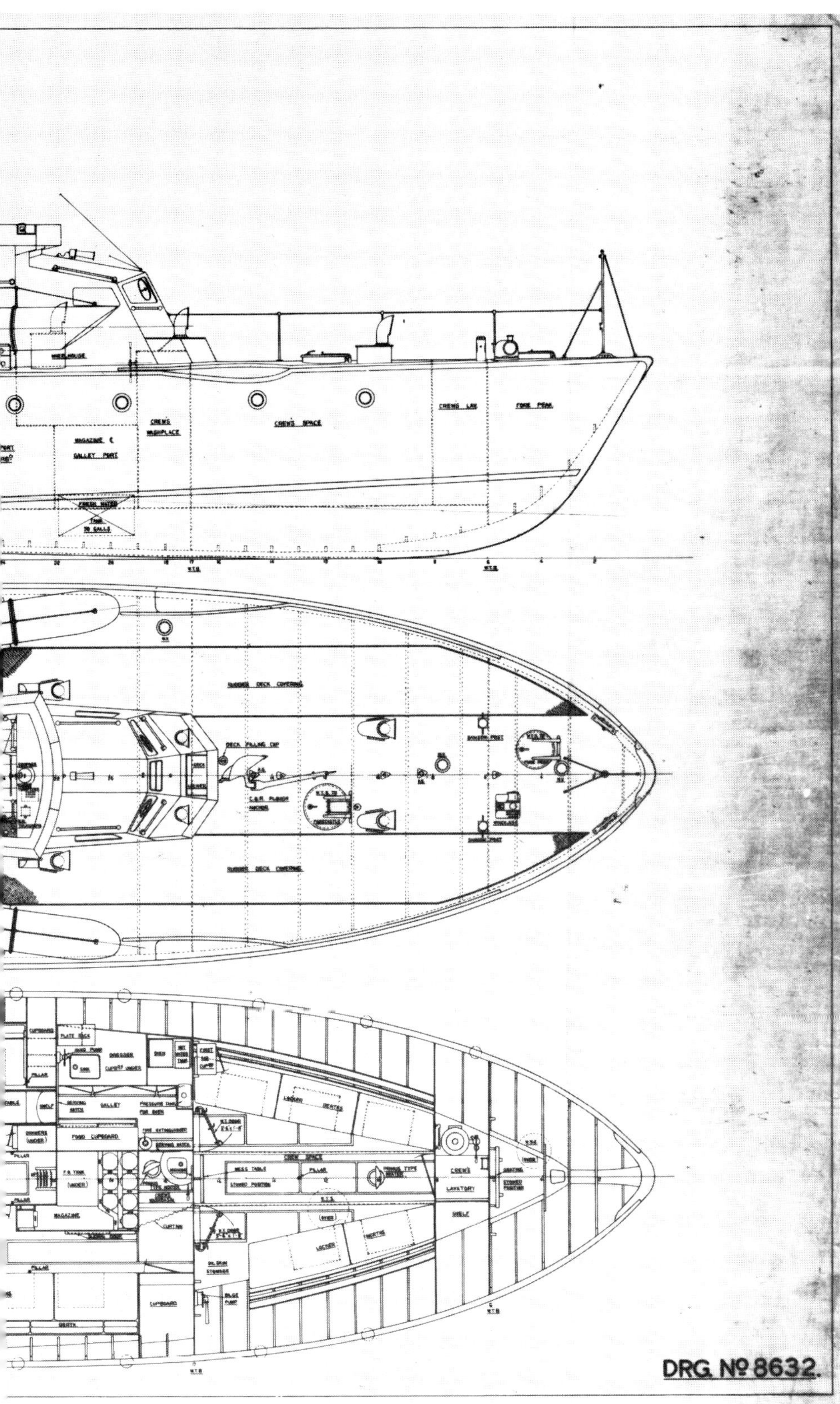

Vosper's first series, *MTB 20-23.*
Vosper

teristics. Du Cane's MTB had concave-convex bottom sections, in which the curve inboard from the chine was dished and the central sections were rounded. The outline was of a yoke, full in the forebody and flattening out on the run aft. (One of the first boats built by Vosper to foreshadow this form was *Swordfish,* an 8·5m [28ft] twin-screw launch built in 1934 and used by the firm for getting about the Solent.) The convex shape provided buoyancy forward and cushioned the tendency to pound when steering into head seas, and provided a useful amount of lift. When, on later MTBs, the forward chine line was raised towards the bow, pounding was even further reduced. This was the principle which had been exploited by Pratt before the First World War (and which he revived in *Tarret*) and even earlier by Saunders with *Napier,* but had lapsed with the post-war popularity of the shallow-V planing hull.

Vosper's MTB was put through heavy weather trials to the satisfaction of naval observers and was bought in October 1937 but was used for six months for further extensive trials before commissioning. In this period, her framing was increased and she was fitted with 20mm Oerlikons in two mountings, the first occasion on which the guns were tried in Britain. The major alteration to the boat was made at *Vernon.* The fore and aft torpedo equipment was removed and a torpedo tube was mounted on deck on either side of the wheelhouse, trained at 10 degrees off the centre line of the boat. It was an arrangement which *Vernon's* staff had for long thought desirable and it set the pattern for all future British, and many American, MTBs. Scallops were cut in the hull and deck at the gunwale to ensure the clear passage of the torpedo, and for three or four years this too became common practice. The boat was highly regarded and in April 1938 was passed into service as *MTB 102.*[8] At this stage the Admiralty committed itself to the Vosper design and proceeded with growing enthusiasm for the 21·3m (70ft) boat, though not its guns. The Oerlikons were removed and replaced by twin 0·5in machine guns in two turrets aft of the deckhouse – this less powerful armament coinciding with the introduction in 1938 of armour plating on S-boats which could withstand 0·303in machine gun fire.

During evaluation with the Royal Navy, Vosper's MTB was fitted with two torpedo tubes mounted on deck, in which form she was commissioned as *MTB 102.*
National Maritime Museum, London

Encouraged by the operation of their MTB, and the necessity of powering future ones, Vosper in January 1938 bought the manufacturing and United Kingdom and Commonwealth sales rights of the Isotta Fraschini, for £15,000. The company was not able to persuade the Admiralty (who at that time could not present a legitimate case to the Treasury for the need for a multitude of MTBs) to establish a factory to build the engines under licence in this country. All Isotta Fraschini engines used in subsequent British MTBs had to be shipped from Italy. In 1938 and 1939 other companies were building MTBs for foreign navies, usually with Isotta Fraschini engines bought from Vosper at £4500 each. Although the seagoing performance of *MTB 102* had altered the Admiralty's view of MTBs from defensive into offensive craft, which might serve on home as well as overseas stations, the anxiety about the absence of a suitable (lightweight, high-speed and available) British engine was not transmuted into remedial action. The Admiralty, like the RAF and the War Office, was traditionally dependent on private industry for financing its own new undertakings and, with nods of encouragement and approval, most companies were shrewd enough to know what the Royal Navy required. For instance, in view of its successful performance it would have been unthinkable for the Admiralty not to have bought Vosper's private venture MTB, though in its endeavours to collect MTB experience impartially some ineffectual boats were purchased.

The Admiralty's powerlessness to foster the establishment of a factory to build Isotta Fraschini engines was no greater an infirmity than its failure to foresee in time the necessity for organised flotillas of MTBs. The essence of the matter was that the suitable engines, the Rolls-Royce Merlin and the Napier Sabre, were aviation engines and, when war came, were required for the RAF. In 1937 the Admiralty did ask Napier if it were possible to develop concurrently a marine version of its 1492kW (2000bhp) Sabre, which was eventually used in a Hawker Typhoon in 1940. As the displacement of MTBs increased under the growing armament requirements, the power of the Napier Lion was seen to have reached a ceiling and the engine's output was soon insufficient. Some hope was attached to a 746kW (1000bhp) V16 diesel engine which was

The engine room, looking forward, of *MTB 102,* showing the 18-cylinder Isotta Fraschinis in the arrangement followed in most of Vosper's MTBs. *Vosper*

being developed for the Admiralty by Davey Paxman and H. R. Ricardo (who had sought the Isotta Fraschini franchise). But Admiralty patronage had been endowed rather late and the first experimental units were not produced until 1939, when they were installed in *Tarret.* Twenty-one Davey Paxman units were installed in some Camper and Nicholson motor gun boats in 1942 but the engines were never in quantity production during the war and afterwards were superseded by the Napier Deltic diesel.

After the acceptance of *MTB 102,* six of this type were ordered in 1938 by the Admiralty; four from Vosper and two from Thornycroft. Also in 1938 the last order for British Power Boat 18·3m (60ft) MTBs was placed. As had happened in the 1880s, Britain's boatyards had full order books for supplying torpedo boats of their own design to many foreign navies. When the Admiralty placed MTB contracts in 1938 and 1939, about forty boats were building for other navies: French, Rumanian, Greek, Siamese, Dutch, Swedish, Finnish, Norwegian, Philippino, and Irish.

The engine problem was solved by Hubert Scott-Paine. Aggrieved at his failure to interest the Admiralty in his 21·3m (70ft) MTB, he shipped the boat to America in 1939 and competed in the US Navy's competitions to find

The Royal Navy's first production MTB from Vosper, *MTB 22* built in 1939. Three more of the 21.3m (70ft) type were sold to Rumania in 1940.

suitable types of MTB, or PT (patrol torpedo) boats. With four years' experience of the problems, Scott-Paine naturally put up a good showing. His boat became *PT 9* and among orders placed by the USN was one with the Electric Boat Company (Elco) at Bayonne, for further quantities of the Scott-Paine type. Scott-Paine negotiated with Packard for suitable engines, the Merlin not being available or even desired by Elco. The size of Scott-Paine's order made it worthwhile for Packard to commence production of their V12 4M-2500 engine. It came to be widely believed in the services in Britain that the Packard installed in British torpedo boats, gun boats and air/sea rescue launches was the Rolls-Royce Merlin built under licence by Packard. This was not the case, though in appearance the V12 supercharged engines were similar. The Merlin was a 27-litre engine but Packard's was 40 litres (2490 cu in) and had a longer pedigree, stretching beyond the invaluable work done by Gar Wood throughout the 1920s and into the 1930s. Later in the war, however, Packard did build the Merlin under licence for use in RAF aircraft.

By 1939 the Admiralty's need for a sea-going MTB had not been fulfilled, and Camper and Nicholson was approached with outline requirements. Camper and Nicholson at Gosport was among the finest of British yacht-builders and its MTB design was based on the lines of a standard 15·2m (50ft) launch which it had been producing since the 1920s – a slim-hulled displacement boat of 20 knots, powered by a 74·6kW (100bhp) Daimler engine. Camper's prototype MTB was a 35·7m (117ft) displacement boat with a firm turn of bilge and fair underwater sections. The beam was 5·9m (19ft 6in) and draught 1·4m (4ft 6in). The planking would be double diagonal mahogany screw fastened to stringers with steel main frames and intermediate ones of rock elm. Three Packards would be installed, and despite the boat's heavy armament and displacement of 96·5 tonnes (95 tons) she would have a maximum speed of 30 knots.

Unfortunately boats of this size and standard could not be built easily in great numbers and the Admiralty was ready to support the scheme proposed in 1939 by Noel Macklin. With his partner, Reid Railton, he ran a motor car business from a workshop in Fairmile Lane, Cobham, Surrey, producing the Invicta and the Railton among other vehicles. The idea occurred to Macklin of using the resources of the scores of boatyards, then without much work, to assemble motor boats from prefabricated parts dispatched from a central production factory. It was a canny notion and he made the most of it by teaming up with a naval architect called Norman Hart, who designed a 33·5m (110ft) hard-chine motor launch. Macklin formed the Fairmile Marine Company in March 1939 and issued a prospectus to interested navies on the rôle of the motor launch in war. The Admiralty was quick to see the possibilities and approved the building of a prototype to Hart's design, specifying the addition of one 3-pounder, two 7·6mm machine guns and twelve depth charges, subject to strengthening the deck and using double instead of single skin bottom planking. The boat was under construction at Woodnutt's yard, Bembridge, Isle of Wight, when war broke out in September 1939. On its trials, the boat achieved 25 knots, powered by three Hall-Scott 447·6kW (600bhp) Defender engines. This sturdy engine (built at Berkeley, California, and developed to power Coastguard revenue cutters to combat the powerful boats of Prohibition rum-runners in the 1920s) had been discovered by Railton when visting America in 1939 to look for a suitable power unit.

The Admiralty ordered a further eleven motor launches, of the type known as Fairmile A, and arrangements were made for the boats to be constructed by ten different firms in the British Isles. Macklin organised the

One of the twelve prefabricated Fairmile A motor gun boats built in 1940 *(ML 103* **was built by Brooke Marine) which were used for escort duties and anti-submarine patrol. In 1942 they were re-armed with a 3-pounder, two Oerlikons (one in place of the funnel) and fitted to carry 9 moored and 6 ground mines for work off the French, Belgian and Dutch coasts.**
Imperial War Museum

despatch to the yards of everything that was required for assembling the boats: fastenings, machinery, stern gear, piping, instruments, deck equipment, wiring, and the timber; plywood frames, sawn to shape and notched to take the stringers; keel, stem and stern pieces cut and scarphed; floors, beams, transoms, planking, bulkheads, interior and superstructure joinery, all were cut, shaped, and annotated for rearing, cuckoo-fashion, in the selected ship and boatyards. All the major prefabrication of timber was carried out in the works of Alfred Lockhart Ltd at Brentford, Middlesex. A shadow factory was established at Arundel, Sussex, which was used principally for the construction of interior joinery. The scheme settled down to commendable efficiency and all twelve Fairmile As were completed in 1940. It was so important a means of production that the Admiralty bought the Fairmile Company in March 1940.

7 FAST FIGHTING BOATS

Once the Admiralty had settled on the Vosper MTB as the basic weapon of Coastal Forces, the aim of the builders was concentrated on increasing the useful armament of the boats without impairing their performance. Vosper's success in this was notable, especially as instructions to fit different guns, increase ammunition storage, or improve the engine silencing arrangements were often made by Admiralty departments unaware of the effect their edicts might have on the fabric of the boats. Maintaining the even trim of fast small craft, to ensure good performance at all speeds, is a skilled business and the balance of planing hulls is easily upset by the addition or alteration of weight in the hull. The problem was severely aggravated by the engine shortage. The 21·3m (70ft) Vosper boat was designed to be powered by three Isotta Fraschini engines, and the propellers were matched to the engine speeds and weights of the boat.

In June 1940, when Italy entered the war, the Milan factory not unnaturally ceased to supply Isotta Fraschini engines to Britain. (When Du Cane was in Italy at the start of the war to place a large order for engines, he was startled to hear the managing director of Isotta Fraschini phone the Bosch factory in Germany and berate them for the late delivery of some generators allocated for the British engines.) With stocks dwindling, eight Vosper MTBs had been fitted with only two Isotta Fraschini engines in order to provide spare parts for other boats equipped with Isotta Fraschini engines. This caused a severe loss of performance in the boats, retarding their operational value. The only engines available in useful numbers during this predicament were American. The Hall-Scott Defender was the 447·6kW (600bhp) V12 marine engine discovered by Reid Railton when searching for a unit for the Fairmile A. It was suitable for motor launches but barely adequate for MTBs. The Admiralty had no choice but to advise Vosper to fit Hall-Scott engines and the half dozen boats built in the second half of 1940 were each powered by three Defenders *(MTB*s *35-40)*. Their poor power/weight ratio made the boats some 15 knots slower than the Isotta Fraschini boats – restricting them to the use of Dover as a base – and the difficulties of installation caused by makeshift adaptation to Isotta Fraschini ancilliary systems resulted in unreliability. Three of these boats were lost in action in 1940, though *MTB 35,* in action in the Straits of Dover on the night of 8-9 September 1941, made clear that intelligence and valour could overcome technical drawbacks. Commanded by Lieutenant-Commander E. N. Pumphrey, this boat sank a German merchant ship in the prototype engagement of MTB warfare. Three MTBs were ordered to intercept a small German convoy, but only two left Dover together, because one boat was unable to put to sea. Pumphrey gently homed in on the convoy, approached his target on the boat's centre engine only, then started the wing engines to approach to 700m (800yd) before firing his torpedoes, one of which misfired and stuck in its tube. He disengaged and was joined by the delayed boat, both being able to attack again later when two motor gun boats arrived and drew the fire of the convoy's escorting E-boats and trawlers. One merchantman was torpedoed.

An MTB's crew looked first for reliability from their boat. High maximum speeds were very fine but if an engine or its equipment failed on the approach to or during an action then that would put the crew at much greater risk than the lack of 5 or 10 knots when disengaging. In 1942, ten Vosper MTBs were fitted with supercharged Hall-Scott engines, of 671·4kW (900bhp), and this slightly improved their performance.

The other engine with which the Royal Navy had to make do was the Sterling Admiral. This had been located by J. S. White, who arranged to build it in the United Kingdom. The Admiralty accepted White's proposals to install the engine in eight MTBs built to Vosper drawings in 1940-1. The engines were supercharged to produce 835·5kW (1120bhp) but they were absurdly heavy for their output and the whole installation weighed some 4·1 tonnes (4 tons) more than if Isotta Fraschini engines had been fitted. In spite of this, the boats had a maximum speed of 39 knots. The hulls were of double diagonal mahogany and plywood was used for the decks and some internal structure, which kept the overall weight fairly light. The boats' displacement was 33·5 tonnes (33 tons) with an armament of two 53·3cm (21in) torpedoes, two 12·7mm (0·5in) and two 7·6mm (0·303in) machine guns. Their endurance was about 30 per cent less than Vosper-built boats and the exceptional difficulties of installing the Sterlings in engine rooms designed for Isotta Fraschinis led to considerable unreliability. This arrangement was not approved by Vosper, who in fact did not know about it, and Du Cane was extremely disturbed that the Admiralty had sanctioned it, not to say aggravated that his company was often disparaged for the boat's performance. Nevertheless, by 1941, MTBs were desperately needed; in war any weapon is better than none. Crews were willing to take out their boats even if their engines were only delivering 60 per cent of their designed power. The five MTBs which tackled the *Scharnhorst* and *Gneisenau* on 12 February 1942, when they were steaming past Dover, were Hall-Scott engined boats; thus the high-speed attackers were hardly the most fleet of predators, though it was the best the Dover base could do. Two boats suffered from engine failure in the engagement and none of them, handicapped by a fresh wind and a bodyguard of S-boats, was able to press home the attack.

White built further boats to the Vosper design, powered by Sterling engines, under a 1940 arrangement by which Vosper released drawings of the 21·3mm (70ft) MTB without payment but in expectation of fees from the use of the Isotta Fraschini engines. With the disruption of the Isotta Fraschini supply, Vosper was left on a sticky wicket until the Admiralty agreed to pay a design and management fee of 5 per cent of the cost of every boat built by other companies. This represented rather more than the fees from Isotta Fraschini installations, but Vosper was intent on insuring against post-war competition from other boatbuilders reaping benefits from its research and development costs. Camper and Nicholson had built two Vosper boats *(MTB*s *29* and *30)* with Isotta Fraschini machinery in 1940 at their Gosport yard, though the company much preferred larger MTBs and at this time was building its prototype of 96·5 tonnes (95 tons) displacement. In 1944 White built six MTBs to its own design, still fitted with Sterlings, but sufficiently improved to maintain the performance of the earlier boats while carrying a greater armament.

The Royal Navy's 1940 estimates for MTB construction called for fifty-five boats to be built in 1941. A new Vosper boat had been designed for this programme, to take the Packard engine, supplies of which were beginning to arrive steadily from America; at the beginning of 1941 sufficient had been delivered to equip two MTB flotillas. The new design was also slightly larger, at 22·1m (72ft 6in), much greater in beam and with an increase in draught of 15·2cm (6in) and an increase in displacement of about 15 per cent. This was due to heavier scantlings, because the 21·3m (70ft) boats were still showing some signs of weakness in rough seas; the fact that the Packard engine, installed in triplicate, was rather heavier than the Isotta Fraschini, though of comparable power/weight ratio; and armour plating on the wheelhouse, the lack of which had resulted in unnecessary casualties from German guns.

Much of Vosper's plant was damaged, as was Portsmouth itself, by air raids in January and March 1941. Four 21·3m (70ft) MTBs had been destroyed on the stocks and, in order to construct the 1941 programme by October of that year (MTB usefulness was limited in the winter months by the poor weather conditions), additional building facilities had to be sought. In 1938 the firm had constructed a new shipyard on reclaimed land in Portsmouth Harbour at Portchester, including a covered fitting-out basin which was invaluable because it enabled work to continue during the black-out. At the Admiralty's suggestion, Vosper also located and constructed a shadow factory at Wivenhoe on the Essex coast.

The combined output of Portchester and Wivenhoe was not sufficient to meet the demands of the 1941 programme and sub-contractors were sought. There were dozens of boatyards in Britain which could have taken on the work but most were producing prefabricated motor launches. Vosper MTBs were not prefabricated and the Director of Naval Equipment, Admiral Sir F. T. B. Tower, pointed out that construction of MTBs differed considerably from an ordinary type of small craft and that it would be 'imperative that the construction by the sub-contractors should come up to the high standard which had been set by Messrs Vosper.'[1] There were six sub-contractors: the Berthon Boat Company of Lymington; Camper and Nicholson, Gosport; Harland & Wolff, Belfast; McGruer, Gareloch; McClean, Glasgow; Morgan Giles, Teignmouth. Between them they built thirty-two boats, under Vosper's supervision, and the remaining twenty-three were built by Vosper's yards. The Admiralty asked Du Cane to visit America in 1941 to liaise with the shipyards who had arranged to build Vosper MTBs in the lease-lend programme. The Admiralty hoped for thirty-two boats, most of which were to be built by Annapolis Yacht Yard in Maryland, but fully understood that the United States Navy might wish to place part of the order with the Electric Boat Company. The first nine American boats shipped to England in 1941-2 were the Elco 21·3m type (70ft), derived from the Scott-Paine boat, and thereafter the fifty-six MTBs, built in American yards, including Herreshoff's, were of the Vosper design. The American contribution is more fully discussed in chapters eight and nine.

During 1942 MTB crews were hard pressed, forced simultaneously to discover and exploit their tactical advantages in action. By this time there was a steady flow of requirements, not by any means uniform, from Coastal Forces operational to material departments. Du Cane occasionally accompanied patrols in the Channel and Vosper made strenuous efforts to produce functional boats. A variety of improvements were made to the boats as a result of consolidating the intensive operational experience of 1941 and 1942, though at

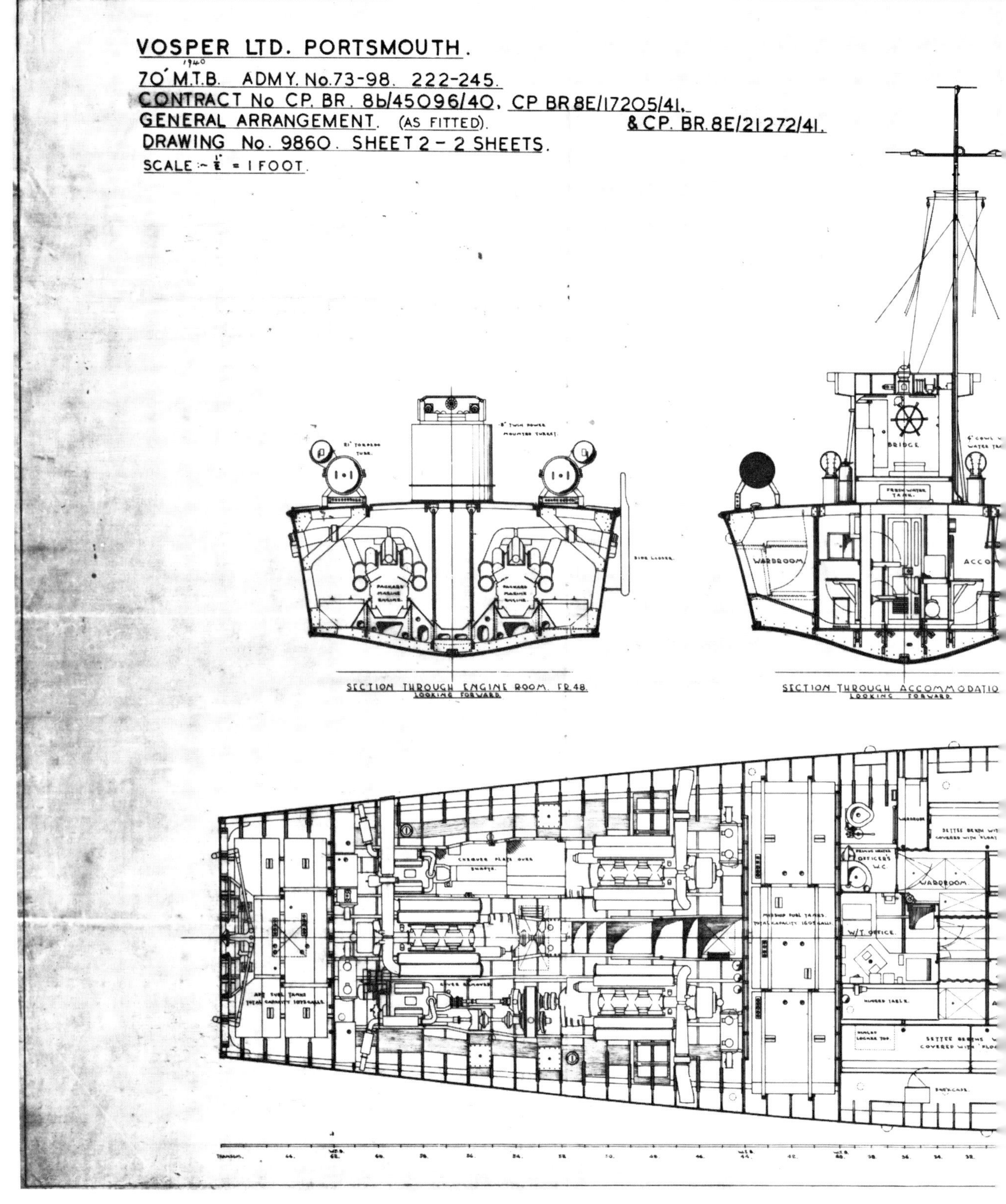
VOSPER LTD. PORTSMOUTH.
1940
70' M.T.B. ADMY. No. 73-98. 222-245.
CONTRACT No CP. BR. 8b/45096/40. CP BR 8E/17205/41.
GENERAL ARRANGEMENT. (AS FITTED).
& C.P. BR. 8E/21272/41.
DRAWING No. 9860. SHEET 2 - 2 SHEETS.
SCALE :- ½" = 1 FOOT.
BRIDGE
WARDROOM
SECTION THROUGH ENGINE ROOM. FR.48.
LOOKING FORWARD
SECTION THROUGH ACCOMMODATIO
LOOKING FORWARD
WARDROOM
W/T OFFICE.

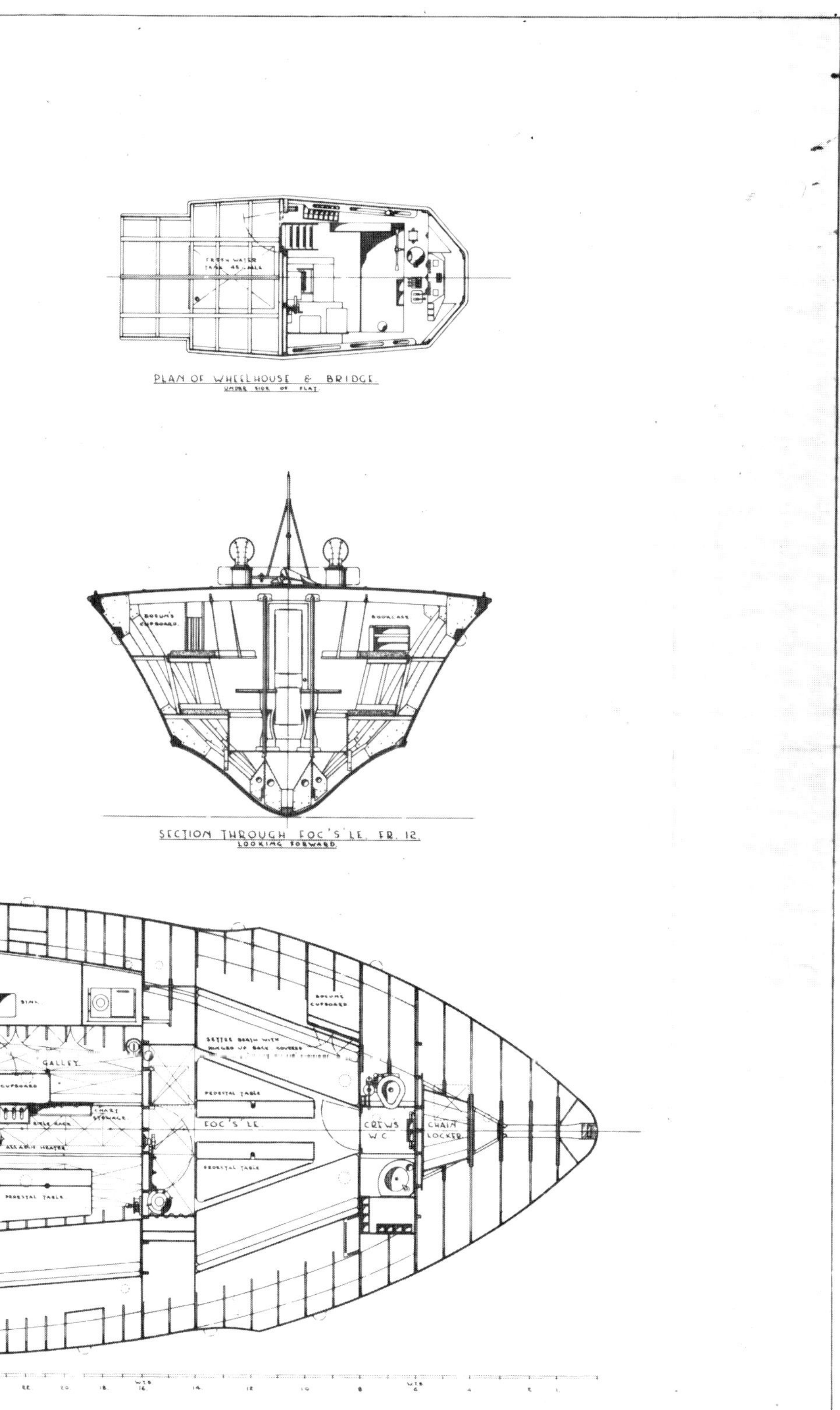

Vosper's 1941-2 22.1m (72ft 6in) series, *MTB 73-98* **and** *MTB 222-245.*
Vosper

times the Admiralty was almost too sensitive to operational needs and bedevilled the builders with contradictory specifications.

It had been found that the 7 or 8 knots produced by the V8 auxiliary engines was insufficient speed for commanders seeking to close the enemy. They needed a greater range of low speeds to suit individual attacks and frequently had to compromise between low-speed silent approaches and faster but noisier attacks, running the central main engine and trusting that a leeward approach would enable the boat to close sufficiently to launch its torpedoes. It was some time before MTBs had to reckon with radar and stealth could still be rewarded. This problem was resolved by omitting the cruising engines (though in some boats a single V8 was installed for manoeuvring and in others a generator was fitted athwartships in the engine room to provide power for the growing electrical demands of MTBs) and directly silencing the main engines. The most effective means would have been to discharge the exhaust underwater, but there was not the time to develop a valve to prevent back pressure. If the outlets were placed ahead of the propellers, the gases emulsified the flow, adding to the hazards of cavitation. The arrangement fitted was the Dumbflow silencer, an expansion chamber heavily injected with water from the exhaust manifold cooling circuit. The exhaust pipes ran extremely hot and had water-cooled jackets for their entire length; the silencers were fitted to the sides of the ship and the water and exhaust were evacuated above water level. This arrangement provided satisfactory silencing up to two-thirds of the engine power, which greatly assisted the prowling tactics of MTBs. It is an interesting mark of S-boats' rôle and affinity with larger ships that their builders were less assiduous in silencing the powerful Daimler-Benz engines; the distinctive bass rumble of the diesels was invariably detected by Coastal Forces crews well before their own boats were heard. It was also possible to smell diesel fumes when nearby S-boats cut their engines to regroup.

Crews learnt fairly quickly the strengths and weaknesses of their boats. In the first two years of fighting, boats often returned to their bases with broken frames and torn planking, usually caused by attempting planing speeds in sea states 4, 5, and even 6. The controlled vigour that marked Coastal Forces crews greatly contributed to their success in action, but the sheer exhilaration of driving a fast boat also led to damaged hulls. This was usually given the blind eye by the staff officers, who recognised that high spirits were fed by high speed and that the war effort was on the whole boosted by unfettered men (though the liberal atmosphere of Coastal Forces bases was continually discomfiting in many Admiralty circles). Propaganda (home morale) departments quickly extracted the romantic flavour of Coastal Forces operations, drawing comparisons with the heroism of the Battle of Britain 'Few' and often reporting outlandish speeds for the boats in impossible sea conditions. Not surprisingly this flavouring was, on occasion, tasted by the young men of Coastal Forces and, while they knew the cold and wet reality of their exploits, it would have been unnatural of them not to play the part assigned by popular imagination. But boats were scarce and no skipper and his crew enjoyed the enforced wait while a boatyard re-fastened a gunwale or the operational base maintenance shop replaced stripped engine-bearings. Eventually the damage caused to hulls by driving at planing speeds into heavy seas was overcome when coxswains learned to tack into the weather, like a yacht sailing to windward, sidling up waves and scooting down them; a judicious hand on the throttles would also prevent the engines racing when the propellers might break the surface as the boat went over the crest of a wave. In anything but light to gentle winds, handling MTBs at high speed called for helmsmanship of the highest order.

This phase coincided with the time when the boats were not immune to structural weaknesses. One of Vosper's major problems was maintaining the homogeneous strength of the hull which they had achieved by 1942 with the necessity to carry a greater payload. Fortunately, useful progress had been made in the manufacture of plywood. Until the introduction of urea-formaldehyde and phenolic adhesives, the veneers of plywood were held together by animal glues. As these are soluble in water, the use of this plywood in boats was limited. Urea-formaldehyde plywoods eventually delaminate if used in a permanently damp environment, as might be found in bilges, but a life of five years could be anticipated if used for bulkheads, superstructures and decks. By the end of the war phenolic resin plywood was used for the topside and bottom skins of some MTB hulls. The use of plywood instead of timber represents an overall saving in weight of about 15 per cent. The direct saving is higher but, while in some multi-stress locations plywood can be used without any doubling-up of the safety factor, when used for decks it was necessary to employ stiffer deck beams than when planks were laid. This particular difficulty was magnified when heavier guns and mountings were added in 1943 and 1944, and steel pillars had to be fitted between the deckhead and floors.

The scallop construction in Vosper's 22.1m (72ft 6in) series. This is *MTB 73*, the first boat to be fitted with Packard engines. *Vosper*

The basic output of the Packard was 820·6kW (1100bhp), but in 1942 a supercharged version producing 1007kW (1350bhp) was introduced. This enabled Vosper to maintain performance as armament, fuel capacity, and equipment were gradually increased but, paradoxically, it led to the inhibition of performance. The diameter and pitch of a propeller has to be nicely matched to the power curves of an engine and, to attain optimum results, screws were designed for boosted engine output. Engines could only be boosted for short periods though, and when running with a full load the engines had to be throttled down to avoid damage from excessive revolutions. This called for sensitive handling from the two or three ratings stationed in the tumult of the engine room and the problem persisted until accurately machined propellers were designed, which married power curves and operating efficiency over a wide range of engine output.

The hazards of propeller cavitation were common to all high-powered boats and generally involved the necessity of reduction gears. There was little accurate data for designers to predict cavitation factors and so integrate with the loss of thrust. Research facilities in the first two years of the war were limited but in 1942 a cavitation tunnel at the Admiralty Experiment Works was completed and this greatly facilitated the study and application of propeller design vis à vis cavitation. Tests made at the tank soon showed that a two-bladed propeller instead of a three-bladed one would increase an MTB's performance by 3½ knots. At first the change-over was accompanied by an increase in vibration but this was later eliminated. The man principally involved in improving propeller performance was the superintendant of the Haslar establishment, Dr R. W. L. Gawn and, in 1944, when many of the more pressing problems had been solved, he organised comparative propeller trials on a Vosper 22·1m (72ft 6in) boat, one of the last to be built during the war. The propellers designed by George Selman, British Power Boat's chief designer, were also evaluated on this boat, *MTB 524,* and Du Cane and Selman cooperated in producing optimum propellers. They also cooperated in tests to find the best rudder designs, but again, at a time when most of the war work had been done and little more could be gained by sanctioning competition between the two firms.

In late 1942 sufficient operational experience had been achieved to make clear that distinctions between torpedo boat and gun boat were superfluous. Arrangements were put in hand to equip the larger gun boats with torpedoes and to increase the firepower of MTBs. The early motor gun boats (MGBs, discussed below) had footling armament and had a tough time when in combat with S-boats in 1941 and 1942 in British convoy routes off the East Coast. If an MGB's prime gun, a 20mm Oerlikon on the aft deck, was knocked out, as happened to Lieutenant E. D. W. Leaf DSC RNVR, in command of *MGB 89* on 29 November 1942, then the boat was endangered and could not rely on holding off an aggressive S-boat with machine guns alone. On that occasion Leaf, in a 23·5m (77ft) Elco boat, had to break off the action and return to Lowestoft. Early in 1943, supplicants from the 5th and 7th MGB flotillas at Lowestoft succeeded in arranging trials to be carried out in an MGB fitted with an Oerlikon on the foredeck in addition to the aft mounting. One of the batch of Elco boats, *MGB 84,* was stiffened forward to take the deck thrust from the gun, and internal fittings, though not to the same weight, were stripped out. The boat's performance was not compromised and thereafter the Lowestoft flotillas were fitted with lightened versions of the Oerlikon.

Other MGBs being built in 1942 and 1943 were designed for and given heavier armament but this experience made it practicable to increase the

The first of Vosper's flush-decked 22.5m (73ft) series ***(MTB 380)*** **of 1944, armed with four 45.7cm (18in) torpedoes and a manually operated 20mm Oerlikon.** *Vosper*

armament of the existing boats. Likewise, *MTB 234* was used for trials in February 1943 to find out if she could function with a 20mm cannon on the foredeck in addition to the one aft. The weight of the gun, mounting and deck reinforcement was more than 1·3 tonnes (1¼ tons), some of which was offset by removing galley equipment and crew lockers. The greater part of this gutting took place in the area beneath the gun, so preserving the trim of the boat. By judicious rearrangement, *MTB 234* was able to maintain her maximum cruising speed of 34 knots and thereafter armament was increased as far as possible wherever practicable. A boat's weapons were often changed throughout her life and by the end of the war boats of the same class invariably had a medley of armament, sometimes 50 per cent greater in weight than when they were built. This was brought about by the changing routine of small boat warfare and by the greatly increased adaptability of the boats as a result of improvements to material, design and construction. In the spring and summer of 1943 many MTBs were given heavier armament and some MGBs were fitted with two 53·3cm (21in) torpedo tubes. A multi-rôle type of boat was emerging, capable of exploiting any circumstances.

Vosper's 1943 class of MTB was a 22·3m (73ft) type of heavier displacement than its predecessors (the prototype was *MTB 379*) and had four torpedo tubes for the old 45·7cm (18in) torpedo. Stocks of the Mark IV and V 53·3cm (21in) torpedo had dwindled by May 1943 and using old supplies of 45·7cm (18in) torpedoes ameliorated the temporary shortage. Fifteen of these boats were built in late 1943 and 1944. Their equipment had reached a luxurious degree: three radar sets, inter-ship radio telephone, and echo sounder, in addition to the hydrophone and wireless telephone set. Heating and cooking by electricity replaced the paraffin-fired apparatus of earlier boats. To save weight, armour plating was fitted only to the bridge and not the wheelhouse as formerly, and the easily ruptured Lockheed hydraulic steering, also prone to leaking, was replaced by Mathway direct linkage gear. Power-operated gun mountings were also vulnerable to gunfire, rendering them immobile, and for the Oerlikons hand-cranking was reintroduced. Du Cane was anxious to raise the chine line forward to reduce the amount of spray which came aboard at sub-planing speeds, but this was disallowed by the Admiralty for it would have meant producing and distributing to dozens of bases a special cradle. The hull form which Du Cane designed in 1937 remained largely unaltered throughout the war. The speed of these boats was less than the 22·1m (72ft 6in) class, but the fourteen 22·3m (73ft)-type of the second series built in late 1944 and 1945 had two 45·7cm (18in) torpedoes and three Packard engines rated at 1119·0kW (1500bhp) and this revitalised their performance, even though they were thick with guns: a twin 20mm Oerlikon aft, twin 0·303in Vickers machine guns on either side of the bridge and a power-operated 6-pounder, forward of the bridge.

Vosper's boats were the orthodox MTBs but during the course of the war, particularly the opening stages, other types were brought into service. A few boats were experimental and some of them, such as White's hydrofoil *(MTB 101)* and Vosper's 13·7m (45ft) hydroplane *(MTB 108)* were lost at sea, while other heralds of hope either failed to reach an operational level or were restricted to target towing and miscellaneous duties. Some of the early British Power Boat MTBs were used eventually for non-combatant purposes. New classes of generally approved types were normally preceded by a prototype which had to undergo exacting acceptance trials before the Admiralty endorsed production of the entire class.

In 1938-9, Thornycroft was building nine CMBs for overseas navies, but

Vosper's experimental 21.3m (70ft) stepped hull, ***MTB 103,*** **shortly before launching in 1940. She was designed for specially powerful Isotta Fraschinis but they were not delivered, two Packards were fitted instead and she was used as a target boat** *(CT 05).*

Vosper

at the outbreak of war the boats were requisitioned by the Admiralty and formed into a flotilla. They were substantially similar to the boats of 1917. Heavier displacement, 17·3 tonnes (17 tons), was overcome at planing speeds by the uprated Thornycroft V12s, of 410·3kW (550bhp), and speeds well over 40 knots were achieved. In addition to two 45·7cm (18in) torpedoes, the boats carried a light armament of twin 0·303in anti-aircraft guns and two depth charges. They plugged an embarrassing hole in British Coastal Forces during the 1940 shortage of boats but the Royal Navy had no intention of concentrating on CMBs. It was galling for Thornycroft, who for sixty years had spearheaded the design and construction of fast torpedo boats, but the CMB was inadequate for the Second World War. The range of these new CMBs was only 200 miles and, though in a light seaway they were the fastest boats in the Royal Navy, very high speed was not a staff requirement and at sub-planing speeds the boats were uneconomic to run. To maintain their high performance, moreover, CMBs had to be kept in a steady state of trim and it was foreseen that the boats would admit of little further development to permit increased payload. Thornycroft refused to fit all the depth charge gear required by the Admiralty because the balance of the boat would have been altered to the detriment of its handling and performance; the boats were less manoeuvrable than the hard-chine stepless hulls. Thornycroft's technical expertise was not defunct; at the same time as *Vernon* was experimenting with forward-firing torpedo tubes, Thornycroft designed and fitted two 45·7cm (18in) lightweight tubes to the 19·8m (65ft) CMB built for the Philippine government in 1938. These tubes were positioned to fire torpedoes ahead of the boat at 10 degrees off centre. Thornycroft also built three 21·9m (72ft) stepless hard-chine MTBs in 1940, derived from a type first built for the government of Eire. They were

Thornycroft built three 21.9m (72ft) MTBs of 40 knots for the Royal Navy in 1939-40. Like six similar boats built for the Irish government they were stepless hard-chine boats, but fitted with three Isotta Fraschini engines of 857.9kW (1150bhp).

sturdy 40-knot boats powered by three 82·1kW (1150bhp) Isotta Fraschini engines. They carried four machine guns and two 53·3cm (21in) torpedoes. In 1941, the third of the trio, *MTB 28,* was lost by fire, doing more damage to its base, HMS *Hornet,* than was ever achieved by enemy action. As can be seen from the table on page 206, loss by fire was the third highest factor in total losses incurred by British Coastal Forces. The number is inflated by the conflagration in Ostend Harbour in February 1945, when a dozen boats were lost, but before the inflammability of petrol is invoked it is worth noting that Italy lost only two boats from accidents with fuel. At sea the use of petrol as a fuel was rarely a unique cause of fire aboard when Coastal Forces were in action. Self-sealing fuel tanks ensured that petrol explosions from enemy gunfire were minimal but of course a fire on board was a severe danger because of the presence of petrol. By contrast, the fuel tanks of S-boats and other diesel powered enemy craft presented a greater danger. Diesel fuel is inflammable and can be ignited by an incendiary shell inside a tank, self-sealing or otherwise, as can petrol, but the vapour in a diesel tank contains a high proportion of air, while in petrol tanks the vaporous nature of the fuel means that very little air is admitted to the tank. Diesel is less volatile than petrol but, once ignited, burns with a greater ferocity, and more S-boats than MTBs or MGBs were lost from shipboard fires as a result of offensive action.

Between the end of 1939 and 1941, Thornycroft also built eight 22·9m (75ft) MTBs, heavy stepless boats of 29 knots. They were powered by four Thornycroft V12s, driving in tandem twin screws. They had a power-operated twin 1·5in gun aft the wheelhouse and two 53·3cm (21in) torpedoes. They were the first MTBs to be equipped with a radar set, but the boats were too sluggish for operational use and, after 1941, all eight were converted into target-towing launches for use by the War Office. In 1940 Thornycroft built four experimental small MTBs *(104-107)* between 12·2m (40ft) and 15·2m (50ft) in length and 9·1-10·1 tonnes (9-10 tons) displacement. They were of stepped design and the first two had marinised Rolls-Royce 820·6kW (1100bhp) Merlin engines, but the idea of carrying them aboard larger ships was abandoned and though the boats were operational no others of their type were ordered. The only other purpose-built MTBs in Coastal Forces were of American origin, though, as discussed below, the influence of Scott-Paine was strong.

In the summer of 1940, German S-boats and R-boats (a *Raumboot* was a 37m [21ft] 20–knot minelayer) started raiding British coastal convoys and laying mines in British coastal waters. This was the S-boats' specific function and they performed it well and with near-impunity at first, although RAF raids and patrols confined German activities to the night-time. However, as with MTBs, darkness suited S-boat warfare and convoys, which had been considered safe at night, were attacked in well-organised raids. In late 1940 one coastal convoy of twenty-five ships lost fourteen of its number between the Thames and Bristol, while many convoys on the east coast lost up to a third of their ships. It was imperative to provide a counter, and before this was found the only defence was to run the convoy westabout round Scotland. The natural enemy of the S-boat was the destroyer, which of course had been especially bred to deal with torpedo boats. Unfortunately, there were few destroyers to combat the S-boats; many had been damaged during the Dunkirk evacuation and the Royal Navy was loathe to submit too many operational ships to the vagarious risks of convoy duties. The convoys were of crucial importance but the shepherding destroyers themselves became alluring targets for S-boats' torpedoes. The few MTBs at HMS *Wasp,* Dover, were hopelessly outmatched in engagements with S-boats and could not be looked to as an effective antidote. Their firepower was inadequate and the likelihood of hitting an S-boat with a torpedo was negligible. Out of this duress was born the realisation of the necessity for a motor gun boat to tackle the S-boat on equal terms.

The only boats available were a few 19·2m (63ft) and 21·3m (70ft) MTBs which British Power Boat was building to Swedish and French orders, and a dozen MASBs under construction for the Royal Navy. The foreign orders were frozen and the boats were completed as gun boats, mounting an irregular assortment of weapons: at first 0·303in Lewis machine gun, shortly replaced by Vickers Mark V 0·5in machine gun. In the first MGBs the machine guns were mounted on the coachroof aft of the bridge but later boats had powered turrets on either side of the bridge. The main armament was a Rolls-Royce 2-pounder gun mounted on the stern deck, which gave way to the 20mm Oerlikon after the summer of 1941 when the specially built factory at Ruislip, Middlesex, started production of the Swiss gun. If the MASBs made poor submarine hunters, they were dismal gun boats. As designed, they were fitted with only two Napier Lion engines, though the pre-war allocation of Rolls-Royce units permitted the Merlin to half a dozen boats. The Napier powered boats had a maximum speed of 23 knots. Nevertheless, gun boats were needed and the completion and conversion of the MASBs was speedily carried out by British Power Boat. When those building with bearers arranged for a twin engine installation were completed, the remaining contracted MASBs were built to take a triple installation. All these boats were substantially like the company's 21·3m (70ft) MTB design, and the triple-screw boats had a maximum speed of 40 knots.

The British Power Boat 21.9m (71ft 9in) MTB/MGB designed by George Selman was probably the best 'short' boat to come out of the Second World War. Built in late 1943 and seen here in Southampton Water, *MTB 459* was one of the boats destroyed in the blaze in Ostend in February 1945.

These improvised MGBs were not satisfactory and arrangements were put in hand for a purpose-built type of gun boat. The most successful of these was the British Power Boat 21·9m (71ft 9in) MGB designed by George Selman. It had an exceptionally strong hull, though a hard-riding one. The tapering superstructure of the converted MGBs was retained but the hollow deck profile aft (which, with the foredeck hump, had earned the boats the soubriquet of whaleback) was levelled and the hump slightly flattened. The first MGB *(74)* was produced in 1942 and nine more soon followed. Three 820·6kW (1100bhp) Packards gave a maximum speed of 40 knots and the armament comprised a power-operated 2-pounder just forward of the superstructure, twin Oerlikons sited aft on the coachroof and twin machine guns to port and starboard amidships, no longer in turrets. As 1007kW (1350bhp) Packards became available, these were fitted to a further sixty-nine MGBs all built by British Power Boat in 1942. This was a remarkable achievement, and though Scott-Paine was by then resident in America, the company continued to run in the manner and spirit instilled by him. More than 1000 staff were employed by this time and the boatbuilding space at the Hythe factory had greatly increased. The boats were built on production lines and the scale of manufacture was second only to the arrangement established by the Fairmile Marine Company. The hulls were planked in double diagonal mahogany and great use was made of prefabricated timbers and plywood to facilitate speedy assembly. Deck and superstructure were plywood and knees, breasthooks, stems, and other structures, which in traditional boatbuilding call for skilled labour, were laminated with pliable strips of timber bonded with glue.

One of the key reasons why Coastal Forces boats could be built in large numbers was the ready availability of timber for boatbuilding. Boatyards' own seasoned stocks did not last for long once they were working on Admiralty contracts, so high was the demand, and consequently many boats were built with green, or unseasoned, timber. This led to the common impression that the boats were expected to last only a few years (a few months, some detractors thought) and this was the case. The Admiralty considered the boats to have a life of eight to ten years. This was not so much because of the use of green wood (the main disadvantage of which was the difficulty of working it compared with seasoned wood and its proneness to rot when used in damp and airless places) but owing to the intensely hard use to which the boats were put. Scores of boats, both British and American built, well outlived their alloted span and many can still be seen as yachts, though more frequently houseboats, around the creeks and harbours of Britain.

The difficulty which was not overcome until well after 1943, was the unreliability of the boats' machinery. This was partly due to an erratic supply of spares, the variety of engines which had to be overhauled, and the exuberant manner in which the boats were sometimes driven. At the end of 1941, directives were issued to improve the reliability of the attack boats. Failure to keep them up to scratch had 'been responsible for the frequent breakdowns, particularly of MTBs and MGBs, which on several occasions have prejudiced important operations'.[2] Half of a flotilla's strength was to be operational at short notice and the remainder was to be undergoing short routine maintenance or long overhauls (changes of engine, slipping, repairing major damage and so on). When a flotilla's strength was other than four, eight, or twelve boats, the benefit of the odd numbers was to be given to the maintenance rather than the operational side. The number of hours an engine could run without overhaul was fixed (as commanders sometimes exceeded manufacturers' recommendations): Napier Sea Lion – 600 hours; Merlin – 350 hours; Isotta

Seven steam gun boats were built in 1941-2 in an effort to provide a sea-keeping deterrent to Germany's S-boats. ***SGB 8*** **(later** ***Grey Wolf*****) was built by Denny.** *Ralston*

Fraschini – 300 hours; Thornycroft V12 – 600 hours; Hall-Scott – 1500 hours (500 supercharged); Packard – 500 hours; Sterling – 500 hours. In spite of their supposed fragility, boats often struggled back to base after withstanding amazing amounts of damage – a tribute to the basic soundness of construction and the seamanship of their crews.

In addition to the short MGBs converted or built in 1940, the design for a more powerful counter to the S-boat was also being sought. The only diesels available then were destined for the large MGBs being designed by Camper and Nicholson and so the Admiralty drew up plans for a large steam gun boat. The Royal Navy's need for a Coastal Forces vessel able to remain operational in rough weather was sufficiently urgent for the liabilities of steam propulsion to be accepted. The light armament and the fragility of the boiler were the prime drawbacks to a potentially effective vessel. The Admiralty designed the lines and general arrangement and worked out the scantlings and details of design with Denny and Yarrow, whose shipyards each built two of the gun boats. The intention had been to build sixty SGBs – steel craft, displacing 172·7 tonnes (170 tons) – but such consumption of metal could not be justified by their military usefulness, and at the end of 1940 orders for just nine SGBs were placed, two of which were cancelled. Hawthorn Leslie built two boats and J. S. White built one. The boats were 44·2m (145ft) in overall length with a beam of 6·1m (20ft) and draught of 1·7m (5ft 6in). They had a single diesel-fired La Mont or Foster Wheeler boiler amidships and two Metropolitan-Vickers steam turbines which produced 5968kW (8000shp). They were very fine vessels, with easily driven hulls, capable of 35 knots. All were launched in 1941, except *SGB 9,* which was completed in February 1942, and though administered by the Coastal Forces organisation their initiation and development was controlled by the Admiralty to a much greater extent than the smaller Coastal Forces boats.

They were far more like destroyers than gun boats and, but for limitations, would have got to grips with armed trawlers and S-boats sooner than they did. Little could be done about their high fuel consumption 3·6 tonnes [3½ tons] per hour at full speed) but early action at close quarters confirmed anxieties about the liability of the boilers to machine gun fire. Their own armament as first designed and fitted was far from menacing: a 2-pounder forward, two twin 0·5in machine guns abreast the bridge and a 2-pounder aft; they also carried two 53·3cm (21in) torpedo tubes. From the middle of 1942 to mid-1943 all the SGBs underwent a lengthy refit except *SGB 7,* which had been lost in June 1942 when stopped by a shot in the boiler room. The major improvement was the cladding of 19·1mm (0·75in) armour plating over the boiler and engine rooms, embracing one-third of the hull area. The armament was also increased and supplemented later, finally comprising a power-operated 6-pounder forward and amidships, a 7·6cm (3in) aft and twin Oerlikons abreast the bridge and on the stern. After the refit SGBs displacement had increased to 264·2 tonnes (260 tons) and the complement raised to 34. Naturally there was a loss of speed, of 5 knots, but the overwhelming firepower was particularly successful against S-boats and confirmed their rôle as hunters – beforehand they were liable to be hunted. They were expensive vessels, in terms of construction, crew and running costs but their performance as seagoing craft was unexcelled in Coastal Forces. This was only to be expected in view of the strength of their construction and the less exhausting demands of seamanship aboard such sizeable vessels; their conception was an indication of the Royal Navy's cyclical habit of expanding the individual size, rather than overall numbers of their attack craft.

However, seven SGBs were not sufficient to satisfy the growing requirements and the jigs which had been used to build the Fairmile A were recovered from storage to produce components for a new boat. This was the Fairmile C, a boat of the same dimensions as the Fairmile A but of 15·2 tonnes (15 tons) greater displacement. Twenty-four boats of this class were built in 1941 as motor gun boats, mounting 2-pounders fore and aft, two twin 0·5in machine guns amidships and two twin 0·303in machine guns on the bridge. Subsequently all weapons but the aft 2-pounder were replaced by 20mm Oerlikons. These MGBs were the first and only boats to be fitted with supercharged Hall-Scott engines, rated at 671·4kW (900bhp), and the first installations used to run very hot until Hall-Scott's engineer seconded to Coastal Forces was able to improve the cooling water circulation. The Fairmile Cs were useful sea-boats and slightly faster than the As, though for either class 20 knots in sea state 3 was a creditable performance. Norman Hart designed both types and, in so doing, played a considerable part in establishing the Fairmile production methods. Because of this he had been led by Fairmile to expect that he would receive some payment for other boats built by the company but this was not immediately forthcoming and after some wrangling the Admiralty made Hart an ex gratia award acknowledging that his work had 'contributed in some measure to the successful construction of boats not to his design'.[3] It was suggested that Hart's contract be severed in October 1940 but the cancellation was not until July 1942.

The low silhouette of the Fairmile C MGB made her a useful convoy escort in the North Sea and the Channel. The boats were similar to Fairmile As but had supercharged Hall-Scotts, slightly better speed and heavier armament.
Imperial War Museum

Between 1940 and 1944, 585 Fairmile B motor launches were produced by 45 companies in the United Kingdom and by more than 30 overseas boatyards. They were round-bilge 34.1m (112ft) hulls of double diagonal mahogany on plywood frames, displacing 66 tonnes (65 tons). They were powered by two 447.6kW (600 bhp) Hall-Scotts and had a top speed of 20 knots and a maximum cruising speed of 16.5 knots. Good sea-boats, they were used for anti-submarine warfare, escort duties, minelaying and minesweeping. Fifty were temporarily fitted with two 53.3cm (21in) torpedo tubes in 1941, another fifty were permanently converted into RMLs for air/sea rescue work with little huts added to the stern deck, and a few others were fitted out as navigation leaders for D Day operations.
Motor Boat and Yachting

The other boats were in particular the Fairmile B, D and H classes. The last was a 31·7m (104ft) troop landing craft. The Fairmile B was a 34·1m (112ft) round bilge motor launch powered by two 447·6kW (600bhp) Hall-Scott engines. It could make 20 knots in good conditions but more importantly it could steam at 15-16 knots in sea states 4 or even 5, though it rolled mercilessly. Its design is usually assigned to the Admiralty, to whom credit is

Q 328

due for resourcefulness but not naval architecture, for the Fairmile B was a rather superior version of the 33·5m (110ft) submarine chaser of which 442 were designed and built in America during the First World War. The Fairmile B was not specifically an attack craft. It was developed mainly for anti-submarine work though fifty boats were ill-advisedly fitted with two 53·3cm (21in) torpedo tubes during 1941. Oerlikons and machine guns were mounted for defence while on minelaying and escort duties, and minesweeping towards the end of the war. Shortly after the start of the Fairmile scheme the Admiralty ordered 145 motor launches, a move reminiscent of the purchase of American MLs in 1915. In fact by 1944, 585 Fairmile Bs had been built, of which 285 were constructed overseas. The Fairmile Bs saw intensive service during the war. Though equipped with asdic, only well trained and efficient crews could achieve successful attacks on submarines and they were most usefully employed in a defensive patrol capacity and in combined operations, where their adaptability and seaworthiness made them invaluable for protecting, navigating and leading in landing craft and the protection of beaches from air and sea attacks.

At the start of the war the Royal Navy's staff requirements for a seagoing MTB had still not been fulfilled and the Admiralty was compelled to draw up lines for a large MTB to its own specifications. This was done and models were tested in the winter of 1939. The hull adopted was a hard-chine semi-planing form, with concave-convex sections for most of the length. The chine line was swept up to the stem head to reduce pounding when driving into head or steep seas and to keep spray to a minimum. The stern sections were wide and almost flat to accommodate the propellers from the installation of four Packard engines and provide a powerful planing surface. When under way, the hull eased itself out of the water on its haunches as the boat's speed increased and rode easily and firmly throughout its speed range. The thinking behind this hull owed something to the work of Gordon Pratt, ably continuing the innovations he showed in *Tarret.*

It was decided that Fairmile should construct this boat because of the large quantities which would be required. This was a bold move and marked the confidence the Admiralty had in the company's capabilities. The boat, which became the Fairmile D, was not only to become a major Coastal Forces weapon but the complexity of its assembly, machinery installation and fitting out was far more complicated than anything undertaken by firms constructing A, B, or C class boats. The D class represented a fairly straightforward progression in design from the C class and its dimensions were similar (35·1m × 6·4m × 1·5m [115ft × 21ft × 5ft]) but its displacement was 92·5 tonnes (91 tons). The first installations of four 932·5kW (1250bhp) Packards was made with direct drive in the absence of gearboxes, producing a speed of 30 knots, but this was increased to 32½ knots (theoretically) when epicyclical 2:1 reduction gearboxes were fitted. The Fairmile Ds were planned as MTBs but the shortage of MGBs and the time saved by not fitting torpedo firing equipment resulted in the first 94 being produced as gun boats. The first Fairmile D *(MGB 601)* was built by Tough Brothers on the Thames at Teddington in February 1942. Tough was the nearest suitable boatyard to Fairmile's headquarters and it built most of the Fairmile prototypes in addition to nine C and B class boats.

Thirty firms in Britain built 229 Fairmile Ds, which were fitted at first with two 53·3cm (21in) torpedo tubes and then four 45·7cm (18in) tubes. The boats with two tubes were classed as MTB/MGBs while the boats with four torpedoes were MTBs only, though all the dually armed boats fulfilled either purpose. By the Royal Navy's own definition of rôles, the MTBs' main function was to 'raid

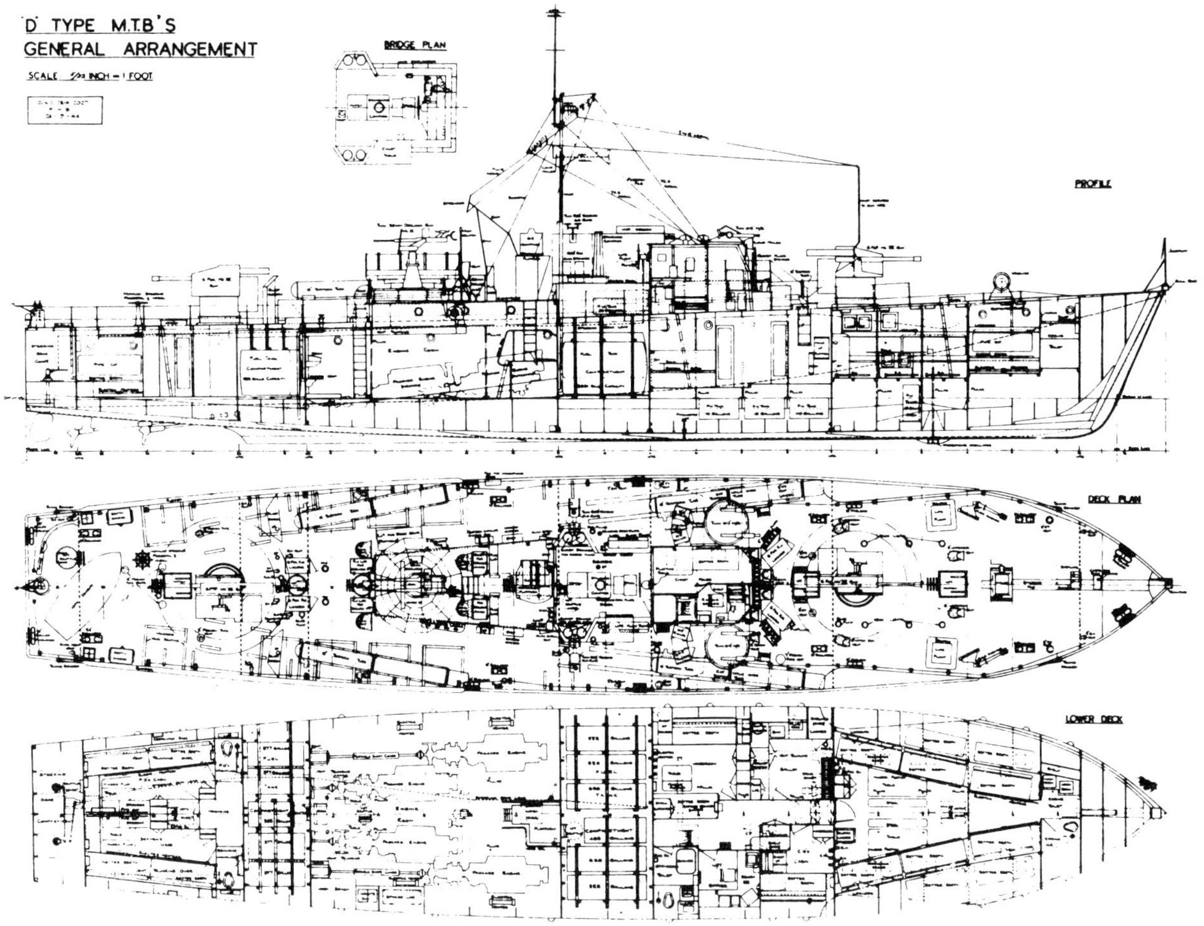

'D' TYPE M.T.B'S
GENERAL ARRANGEMENT

BRIDGE PLAN

PROFILE

DECK PLAN

LOWER DECK

the enemy's shipping in his coastal waters and harbours, relying upon surprise and relatively high speed to get within striking distance'.[4] The range of the D-class MTBs, 1200 miles at 10 knots, enabled torpedo attacks to be made in enemy waters and, from late 1942 onwards, a flotilla of such boats based at Lerwick carried out many successful torpedo attacks on the Norwegian coast. When the D-class boats were equipped with torpedoes they met the operational requirement of filibustering MGB skippers who were frequently powerless to torpedo enemy shipping they encountered. From 1943 onwards, all the Fairmile Ds were fitted as combined MGB/MTBs but the nature of the boat meant that, in home waters, they fulfilled primarily the rôle of gun boat, whose main function was to 'prevent enemy Coastal Force craft from raiding [Allied] shipping, relying upon speed and manoeuvrability to catch the enemy and upon superior fire power to fight him'.[5] The subsidiary function of gun boats was 'to aid MTBs in attacks on enemy shipping by engaging escorts and creating diversions; in Combined Operations to engage shore targets; and to destroy small coastal shipping by gunfire alone'.[6] In these capacities the Fairmile Ds saw furious service in the Mediterranean, though the circumstances of sea warfare there offered wider employment of their dual rôles.

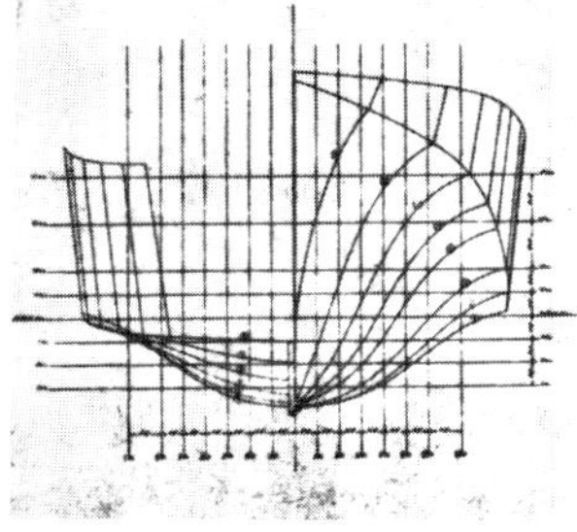

Plans of the 1944 Fairmile D MTB/MGB. *National Maritime Museum, London*

Because of their size, relatively high degree of seaworthiness, and deceptively comfortable ride, the Fairmile Ds were often over-driven in rough weather, as the 21·3m (70ft) MTBs had been in 1939-40. This 'brought to light some of the limitations of the Fairmile method of construction'.[7] Under normal circumstances such large boats would have been built in the conventional manner, as were the Camper and Nicholson MGBs, but mass production was the only means of satisfying the need for large numbers. The forward plywood frames were prone to breaking and the glued scarphs on the keels were also liable to fracture. Sufficiently long runs of timber were increasingly difficult to find and the Lockhart works had to glue scarph joints for gunwales and stringers. The resins had to be cured at warm temperatures and quality control was difficult to maintain. Nothing could be done about the fracturing of the scarphs except to reinforce them with steel fishplates and the number of frames in the forward sections was doubled. Great care was taken over the design of the boat, but once laid off and the templates prepared, the Ds and all Fairmiles went straight into full production, when there were severe limitations to the changes which could be made. All modifications, and there were many, had to be effected within the existing structure of the design. It would have been impracticable to alter any of the jigs because of the long chain of time-consuming repercussions and the designers at Fairmile were creative and dexterous in finding solutions to the flow of problems. The hulls of the Fairmile Ds were double diagonal mahogany but the decks and superstructure were of plywood. As designed the deck abaft the bridge was sufficiently strong but, during one of the periodic rounds of re-arming, Oerlikons were re-sited on the coachroof over the engine room and a 6-pounder was positioned aft, subjecting the decking to impossible strains. Fairmile designed and distributed a kit of supplementary parts to reinforce the large expanse of tremulous deck but many flotillas, particularly those in the Mediterranean, had to cobble strengthening members from the meagre sources of their own bases until the approved parts arrived.

Although the construction was largely a matter of structural assembly the amount of plain boatbuilding which remained was considerable. The Admiralty supplied, through Fairmile, items such as w/t gear, asdic, signalling flags, lifeboat, and lifejackets, and Aldis lamp and the boatyards provided and fitted the boat's dinghy, anchor and chain, navigation lights, stores, and so on. All machinery, controls, wiring harnesses and armament were supplied through Fairmile, and it was in the installation of the four Packard engines and two electricity generators with their attendent fuel and cooling systems, electrical screening, the fire extinguishing equipment, and the increasing complexities of

A Fairmile D under construction at Kris Cruisers yard on the Thames at Isleworth in late 1942. The notches and slots in the prefabricated plywood frames give an idea of the relative simplicity with which these 96.5-tonne (95-ton) craft were constructed.

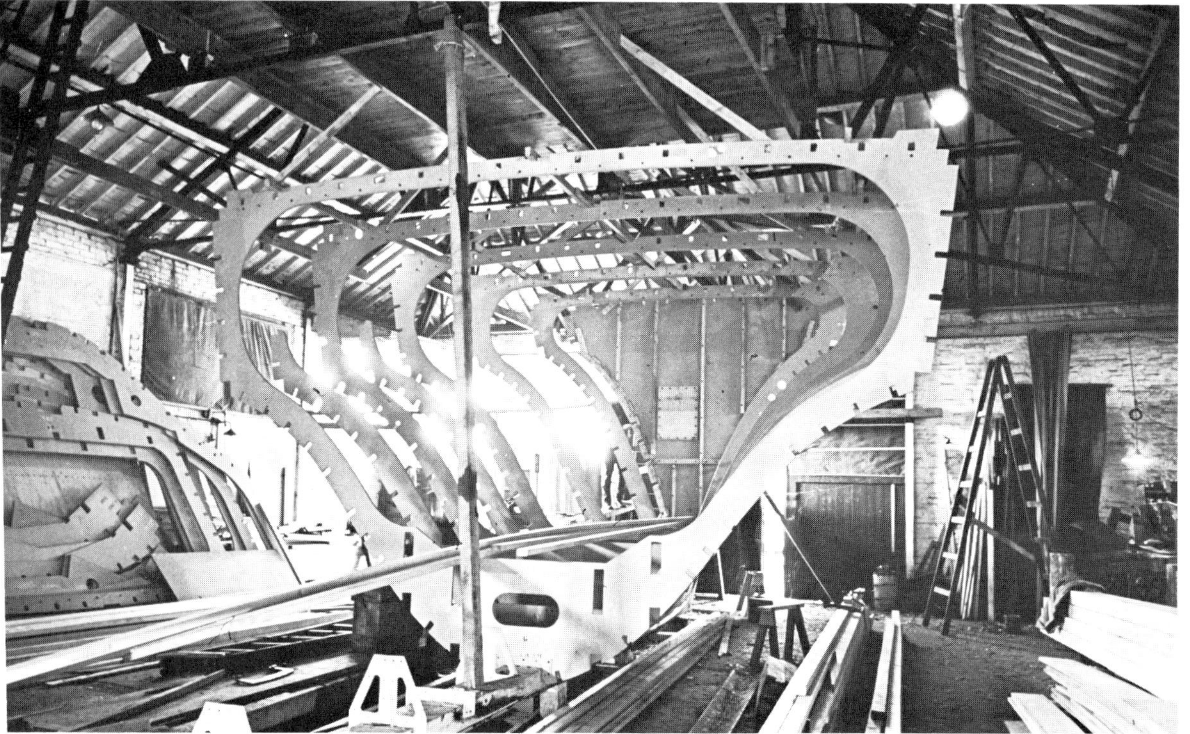

A Fairmile D, *MTB 747*, disposed of by the Admiralty at a yard on the Upper Thames at the end of the war. The first 32 Fairmile Ds were completed as MGBs with flush decks, thereafter scallops for two 53.3cm (21in) torpedoes were incorporated.

the control and instrumentation that the selected boatyards proved their worth. With such an important boat as the Fairmile D, the yards were selected with care and though there were some niggling mishaps (such as one yard which used material sufficient for six boats to build just one) these were easily outweighed by the rapid speed and sound quality of production and the immunity from bomb damage as a result of the yards being scattered about the country. As an acknowledgement of the sprouting complexity of construction, the builder's profit allowance on Fairmile Ds was increased in 1943 from £325 to £375 per boat (with effect from *MTB 724*). The average cost of each boat increased by 15 per cent; the Fairmile MLs remained at a steady price and were half the cost of the Fairmile Ds, or Dog boats as they were widely known. Yards building Fairmile Ds which were consistently cheaper than average were generally 'those with pre-war experience of yacht building and especially where the directors or proprietors are practical craftsmen and have worked with and exercised direct supervision of their employees'.[8] Because of the continual changes to specification of the Fairmile D the high increase in wages during the war was not offset by the factor of repetition which was so beneficial in Fairmile B production.

The most expensive Fairmile Ds were on the whole those built in some London and Southampton yards; the most reasonable and officially favoured, those built in Scotland, Ireland, and at some south coast builders. The costliest Fairmile was the experimental F type which was built by Kris Cruisers, of Isleworth on the Thames, for £12,188. (These prices are only those charged by the builder and do not represent the full cost of material because the Fairmile organisation was owned by the Admiralty.) The Fairmile F, or *MGB 2001*, was basically a D-type hull in which it was proposed to install air-cooled aviation engines. Much of the boat's inception and development was by Fred Cooper, who had joined the Fairmile design team late in 1941. The aim was to produce a high-speed MGB with readily obtained powerful petrol engines. This ruled out the water cooled variety, notably the Rolls-Royce Merlin and Griffon, the Napier Sabre, and the American 24-cylinder Allison W3420 engine. Of the residual air-cooled radials, the most powerful British engine was the 14-cylinder 38·7-litre Bristol Hercules which developed 1230·9kW (1650bhp). Cooper fitted four Hercules engines in this prototype vessel, experiencing immense difficulties with the installation and with arranging adequate spray-proof air ducting. This latter problem was to trouble Vosper when fitting gas turbine machinery in MTBs after the war. The Fairmile F eventually ran trials

The only known photo of the experimental Fairmile F, ***MGB 2001,*** **seen here under construction at Kris Cruisers in 1942. She was the most powerful petrol engine boat built in Britain and one of the most expensive light craft produced during the war.**
Motor Boat and Yachting

in late 1943, and logged a speed of 36 knots, creditable for a boat displacing almost 101·6 tonnes (100 tons), but *MGB 2001* was never commissioned and the bold experiments with engines and her variable-pitch propellers were overtaken after the war by the appearance of the excellent Napier Deltic diesel.

Probably the finest boats of Coastal Forces were those designed and built by Camper and Nicholson and based on *MGB 501,* designed in 1939 as an MTB but built and completed in February 1942 as an MGB. Camper and Nicholson were not willing to involve themselves in the Fairmile scheme and though they built six Vosper MTBs under licence in 1940 and 1942, much of the firm's war work for the Admiralty, built at the Gosport yard, was of a special and individual nature. The prototype and the eight which followed were displacement hulls of 96·5 tonnes (95 tons) (35·7m × 6·1m × 1·4m [117ft × 20ft × 4ft 6in]) built of double diagonal mahogany on steel frames, with intermediate pairs of Canadian rock elm frames. The planking was screw-fastened to timber stringers outside the framing, resulting in a lightweight and exceptionally strong hull. The prototype had three 932·5kW (1250bhp) Packard engines but the remainder (apart from the last, *MGB 509*) were fitted with three 746kW (1000bhp) Paxman diesels. The prototype was lost in action in the same year she was launched but the rest of the boats on the stocks, built ostensibly for the Turkish government, were taken over by the Admiralty. The first two boats, *MGB*s *502* and *503,* were fitted as gun boats, armed with a 2-pounder aft, twin 0·5in machine guns abreast the bridge and two twin 0·303in machine guns amidships. They carried two 53·3cm (21in) torpedos. The next five boats were converted during the course of construction into cargo boats for ferrying machine tools and ball bearings from Sweden via the Skagerrak to Hull. A small hold forward and a larger hold aft replaced the accommodation and most of the crew were housed in the superstructure erected aft to give the boats the appearance of small coasters. Loaded with 45·7 tonnes (45 tons) of cargo, the boats could make the hazardous passage at 20 knots. From 1943, when the blockade-running flotilla started, until the end of 1944, the five boats bore names: *Hopewell, Nonsuch, Gay Viking, Gay Corsair, Master Standfast.* Two of the boats were lost while on these operations. Because of the scarcity of Paxman diesels the last boat in the 1942 series, *MGB 509,* was fitted with three Packards and had a maximum speed of 31 knots, which was a little faster than the diesel powered boats. In 1947 this boat, renumbered *MGB 2009,* became the first to be propelled by gas turbine machinery when trials were carried out with a Metropolitan-Vickers 1865kW (2500shp) Beryl engine, driving the central shaft only.

In 1944, another eight MGBs were built by Camper and Nicholson, powered by three Packard engines. The armament of this group, and the remaining boats of the first flotilla, was increased to a 6-pounder fore and aft, Oerlikons amidships and abreast the bridge, and four 45·7cm (18in) torpedo tubes were fitted. The second series boats were 0·6m (2ft) wider in beam and 20·3 tonnes (20 tons) greater in displacement, with a corresponding decrease of maximum cruising speed to 26 knots. Both flotillas were splendid sea boats but the only active employment for the second batch was escort duty to and from Antwerp.

Notwithstanding the Royal Navy's perilously late start in MTB, MGB, and ML design and production, the woeful lack of engines, the disgraceful evaporation of First World War experience, and the absence of an authoritative central organisation, the men and material of what was to become Coastal Forces put up a remarkable performance in the early stages of the war. In addition to their enemy at sea, the personnel had to combat at home some official disdain and disinterest, pitiful shortages of spare parts and maintenance facilities and what appeared to be abnegation of any responsibility for tactical advice and information. That Coastal Forces won most of their battles and their war is due to the indefatigability and spirit of the personnel. It was a prerequisite of early operational service in Coastal Forces that personnel had experience of handling fast boats – a qualification that hundreds of men were peculiarly able to satisfy. In the early days of Coastal Forces there was a leavening of Royal Navy officers and ratings among the large intake of RNVR and Hostilities Only crews and most of the volunteer officers had, if not fast boat experience, at least powered boat or sailing experience. Without this feeling for small boats and their way in a sea, without the rudiments of seamanship they teach, it is unlikely that enthusiasm alone would have given Coastal Forces crews the victories they worked so hard to achieve.

The failures in Coastal Forces were of an official nature. There was in many quarters a strong antipathy towards the capricious young men who served in MTBs and the like. The appointment of Rear-Admiral Piers Kekewich in November 1940 as Rear-Admiral, Coastal Forces, was belated, and the post's responsibilities were chiefly administrative, training, design, supply, repair and maintenance. Coastal Forces headquarters was in Portland. The operational aspect, and consequently much of the design matters, were handled at the Admiralty by the department of the Deputy Director of Operations (Coastal), established in July 1941. There was much muddle and duplication of work by the two organisations but the main problem was that RACF did not have the authority to implement the responsibilities he had been given. Gradually Kekewich's responsibilities were disseminated until by August 1941 his chief staff officer, Captain Augustus Agar, noted that they 'were left with little to do at Portland except look at the Bill from our office windows'.[9] Of course by that time Coastal Forces had been set on its feet: MGB construction programmes put in hand; bases set up around the coast; the training establishment at Fort William and the working-up base at Weymouth commissioned. In his reports recommending the dissolution of the post of RACF the Deputy First Sea Lord, Admiral Sir Charles Kennedy-Purvis, KCB, acknowledged that:

'the institution of RACF's organisation two and a half years ago has fully justified itself. The co-ordination of all Coastal Force work has been well done. The Coastal Force had an erratic start and it badly needed some central authority to pull it together. Now that it has been pulled together, I consider that it is time for the Force to be treated in the normal manner and be absorbed entirely into the Admiralty and Commander-in-Chiefs' organisations.'[10]

The re-organisation took effect from 12 February 1943. The catalyst for the change was the rapid increase in boats and personnel which had become apparent by 1942, when it was finally realised that the size, though not the importance, of the Coastal Forces warranted treating it as a functional service much like the Fleet Air Arm or submarine service. The important practical improvement arising from the reorganisation was the rationalisation of the Coastal Force supply depot at Fareham, Hampshire; the poor distribution of spares and equipment had been a supply officer's nightmare. The engine overhaul shops at Shepton Mallet, Somerset and HMS *Hornet,* Portsmouth Harbour, also benefited. The efficient running of HMS *Hornet,* a branch of HMS *Vernon* until 1940, was a vital spark in the early life of Coastal Forces.

Evidence of undisciplined policy was apparent, before and after the re-organisation, in the sundry boats ordered for Coastal Forces. Experiments in design are valuable but experiments in purpose are expensive. Once the dual rôle high-speed MTB/MGB was found to be the craft best suited to Britain's needs then all developments should have been focused on that type. That the necessity existed in the middle of the war to order, for experimental purposes, such boats as the Fairmile F or the 30·4m (100ft) Vosper *(MGB 510)* is an indication of giddiness. For such reasons it would seem desirable to have maintained a nucleus of Admiralty staff to determine whether the Royal Navy needed a few dozen long-term boats, like the excellent but expensive Camper and Nicholson type of MGB/MTBs, or hundreds of cheaper and quickly built 'throwaway' boats. The first option ensures a trained body of men and an operational support service; the second presupposes a certain readiness to expend men as well as boats, and depends on the experience and ability of British and Allied companies to provide large quantities of engines and craft. Fortunately in 1939 the enormous support necessary for the second choice was forthcoming.

A pair of the numerous 20.7m (68ft) ('Hants and Dorset') ASRLs built by British Power Boat between 1942 and 1944, seen here at St Tropez in 1944. During the war ASRLs saved 5721 Allied airmen in British waters; 3200 airmen in foreign waters, and 4665 sailors, soldiers and civilians in foreign waters. *W. Onions*

In October 1941 the Admiralty proposed an amalgamation of its rescue services with the RAF. The Air Ministry declined the offer of its building, maintenance, manning and training services being jointly administered with the Navy but, at a meeting in March 1942, arrangements were made for pooling resources and by July 1942 the Navy and RAF were sharing their bases for mutual assistance in repair and maintenance of air/sea rescue launches.

Hundreds of ASRLs were built for the RAF during the war. At first they were not armed; it was hoped that their rôle of rescuing ditched airmen would immunise them from attack. Because of the similarity of their appearance with offensive boats (the hulls and machinery were alike) this could not be relied upon, and consequently they were armed with anti-aircraft guns. Most boats were provided by British Power Boat and Thornycroft, who had previous experience of RAF requirements, but Vosper also designed and built a few ASRLs for a special requirement. Seaworthiness was as important as high speed because, unlike MTBs and MGBs, ASRLs could not be choosy about the state of the weather when they put out to sea. The ASRLs saw active service in every part of the world and rescued thousands of airmen (including Axis aircrew). In 1944 the RAF placed orders for long-range Air/Sea Rescue Craft for use in the Pacific. Two craft of almost wondrous construction were laid down at Camper and Nicholson, to be powered by four Rolls-Royce V8 1305·5kW (1750bhp) Griffon petrol engines, but the boats remained uncompleted when the war ended. British Power Boat built the hull of its 33·5m (110ft) version but work again stopped in 1945 before the boat could be fitted out and, in 1946, the entire boatyard at Hythe closed down. Vosper also drew up plans for a 35·1m (115ft) boat. The only boat which was completed, in September 1946, was the 32·0m (105ft)-type from Kris Cruisers, designed by Fred Cooper and, like the Fairmile F, powered by four Hercules engines.

8 EARLY EXPECTATIONS

In September 1939 all these developments were hidden in the future. Germany had some twenty-one S-boats available for service, Poland had ordered two J. S. White motor anti-submarine boats (MASBs) from Britain, and France was showing interest in eighteen 21·3m (70ft) MASBs from British Power Boat. French First World War light craft had been of an anti-submarine type. Such French boats as existed between the wars were classed as *vedettes torpilleurs* A and B (patrol torpedo-craft, small and large), *VTB 8, VTB 10, VTB 11* and *VTB 12* being in service at the beginning of the Second World War. Measuring 18·9m × 4·2m × 1·2m (61ft 7in × 13ft 8in × 3ft 11in), they displaced 28·4 tonnes (28 tons) and were powered by two 820·6kW (1100bhp) petrol engines, giving a speed of 45 knots. Their armament comprised two 45·7cm (18in) torpedo tubes and two machine guns. *VTB 9* had been lost after being damaged at high speed in rough weather off Cape Barfleur in August 1939. Sister ships were under construction at Meulan.

Poland's other ally, Great Britain, had three flotillas of MTBs and the 1st MASB Flotilla: the 1st MTB Flotilla *(MTB 1* to *MTB 6)* had been in Malta since June 1937; the 2nd *(MTB 7* to *MTB 12)* had been shipped to Hong Kong in 1938; and the 3rd *(MTB 14* to *MTB 19),* destined for Singapore, was still passing through the Mediterranean as the Second World War approached, and so was held there. These British Power Boat 18·3m (60ft) MTBs, together with their near-sisters the MASBs, provided the only experience upon which future operational plans could be based.

Briefly, their main functions would be the Royal Navy's traditional ones: to protect friendly coasts, lines of supply, and communications; to attack the enemy's coasts, lines of supply, and communications; and to cooperate with the other services. But unlike First World War CMBs and Germany's S-boats, MTBs had not yet been allocated any special function of their own. Nevertheless MTBs at speed did look spectacular, and some of their pre-war protagonists claimed great things for them. They would be able to do 40 knots and have a range of 500 miles at 20 knots. Being capable of operating in 4-6m (12-20ft) seas, they were ideally suited for attack at night, in low daylight visibility, and in heavy weather whatever the visibility. In fact, the rougher the better, for then the MTBs would be almost invisible in the troughs and their violent motion would make them impossible targets, while the enemy battleships would stand out as rock-steady aiming points. Thirty to forty MTBs all manoeuvring together at night in bad weather would have a demoralising effect on a battleship's crew. At dawn, the capital ships would be lying helpless to be finished off by aircraft bombs or battleship gunfire. Avoiding every tiny port or inlet which might conceal a swarm of mosquito craft would impose a severe handicap on naval strategy, and, if the hostile battlefleet declined to come within a 200-mile radius, then the MTBs could be carried to sea in a parent ship's davits and released to creep into the enemy's lair, inflict mortal wounds, and escape at high speed.

That was the theory of fast fighting boat warfare, but during the last few months of peace there were indications that it was a somewhat imaginative theory. Admiral Sir Dudley Pound, Commander-in-Chief Mediterranean, considered that attacks on harbours 'were perhaps the employment of Motor Torpedo Boats which is most frequently visualised',[1] but foresaw that the likelihood of important enemy ships lying in mildly defended Mediterranean ports was so remote that this function should not be taken into consideration.

The 18·3m (60ft) British Power Boat craft were also the smallest operational MTBs (as opposed to CMBs) to serve in the Royal Navy, their size the result of stipulation that they should be transportable by rail. Furthermore, they had the most elementary planing sections, later boats having greater deadrise to be more sea-kindly in rough weather. They were certainly seaworthy, but their normal continuous speed was only 29 knots, with a maximum of 33 knots. When trying to plane in a seaway, they climbed clear of the wave and buried their bows going down the other side. The continual jarring caused by driving hard in such conditions exhausted their crews and turned even the simplest of shipboard tasks into an arduous struggle: the boats themselves started to suffer frame damage and their side planking seam strips worked loose; the aluminium deck corroded and had to be replaced by double diagonal mahogany planking. Between January 1938 and March 1939, each boat spent an average of seventy-six days refitting and twenty-seven days under repair in Malta Dockyard.

It was very quickly learned that much damage could be avoided by experienced use of helm and throttle. Even in calm waters at night, high speed was not always advantageous, for then the high pluming wake could be seen for miles. Neither gunnery nor torpedo launching arrangements were very satisfactory. The machine guns could be resited on ring supports in armoured turrets on either side of the deckhouse but nothing could be done about the torpedoes.

The penalty for years of neglect of MTBs was considerable. Not only was the Royal Navy unsure of what MTBs were supposed to do, but the very cost of them not even doing it was high in personnel and shore facilities. The MTBs were small, but each boat's complement included several skilled ratings. They were subjected to great strain, although in fair weather the crews could stay at sea for up to two days before fatigue set in. The boats were theoretically independent of the shore but crews rarely slept on board when in port, although the Hong Kong flotilla once had to endure cramped companionship for some months when their billets were untenable.

A slipway or crane was essential for repair and overhaul, special slings and cradles also being necessary, as was an engine workshop. An MTB might have been built largely by carpenters, but only shipwrights could carry out acceptable repairs. But no amount of zeal could make the boats serve in an offensive rôle before the war. They were gradually relegated to local defence and patrol duties, and as a result an atmosphere of anti-climax followed their introduction to the Mediterranean.

Yet in spite of these apparent disappointments, the British Power Boat 18·3m (60ft) MTBs did give an indication that such craft had some future

In three years the equipment aboard fast fighting boats grew from the inadequate *(MTB 83,* **above) to the point where commanders were complaining about the excess of systems on the bridge** *(MTB 523,* **below). The all-important boost gauges are above the throttles and to the right of the compasses are the tachometers.** *Vosper*

value. By September 1939 the Hong Kong boats had been joined by two Thornycroft 16·8m (55ft) single-step CMBs originally built for the Chinese Navy. A flotilla, which was to become the 4th MTB Flotilla, was gathering at Portsmouth. So far, it consisted of *MTB 40K,* 20m (65·4ft) long and the sole representative of Aero-Marine Engines Ltd; *MTB 100,* converted from a fast motor minesweeper of British Power Boat 18·3m (60ft) design; Vosper's *MTB 102,* now repaired after damaging her side planking when being driven hard off the Needles; and *MTB 22.* This last was the first of the Vosper 21·3m (70ft) boats, three sisters originally ordered for the Royal Navy being sold to Rumania and christened *Viforul, Viscolul* and *Vigelia.* Several Royal Navy officers present at the ceremony laid 10s (50p) bets with each other on whether these boats' torpedoes would eventually be used against Allied targets.

Experimental craft were also being designed or were under construction: J. S. White's hydrofoil *MTB 101; MTB 103,* a Vosper 21·3m (70ft) stepped boat; *MTB 104,* a 15·2m (50ft) Thornycroft; *MTB 105* and *MTB 106,* 13·9m (45ft 9in) Thornycrofts; *MTB 107,* a 12·5m (44ft 3in) Thornycroft; *MTB 108,* a Vosper 13·7m (45ft) hydroplane; *MTB 109,* a 12·5m (41ft) semi-hydrofoil built by Denny; Vosper's *Bloodhound;* and *Tarret.* In addition to these vessels on order for the Royal Navy, fast fighting boats were being built in British yards for foreign customers with the Admiralty keeping a watchful eye on their progress.

Germany's total of twenty-one S-boats was thus matched by twenty-nine MTBs and MASBs in Royal Navy service, but the British boats were scattered across the world and were of seven different types. The reason for this diversity was later explained by Admiral Sir William James, then Commander-in-Chief, Portsmouth. Little money was available for prewar construction of light coastal forces and it was decided to invest it in prototypes. It would have been unwise to build a great fleet of high-speed craft only to find them redundant, worn out, or obsolete when a crisis occurred in a year or so. If war did come, the lessons learned could be incorporated in the accelerated production of the most suitable type. What was serious was the absence of a training establishment and of senior officers with experience of such boats to administer and then instruct the younger enthusiasts. But as this was true of most navies, it was perhaps not such a disabling handicap.

One nation that did have modern experience was Italy, whose MAS-boat crews had practised the silent stalking attack and only used high speed when withdrawing after launching their torpedoes. Some of the Baglietto 16m (52ft 3in) 11·8-tonne (12-ton) type *(MAS 204, MAS 206, MAS 210, MAS 213* and *MAS 216)* were still serving in 1939, although they had been built in 1918 and were now only capable of 10-15 knots for about an hour. The other First World War and interwar boats had been replaced as they wore out, so the Royal Italian Navy had continued fast fighting boat development during peacetime. In addition, Italy had the Isotta Fraschini Company who were maintaining the production of large-bore, short-stroke, medium revolution petrol engines, so suitable for high speed craft. Apart from the aircraft business, British manufacturers tended to concentrate on engines with a small bore, a long stroke, and high revolutions, a design imposed by British car tax regulations.

Two Isotta Fraschini 559·5kW (750bhp) petrol engines were installed in *MAS 423, MAS 424* and *MAS 426,* which had been laid down in 1923 but not completed until 1928-9. Basically the same as the earlier SVAN 16m (52ft 3in) 12·2-tonne (12-ton) type, their speed was raised from 25 to 40 knots by these new engines, while still mounting two torpedoes, two machine-guns and depth-charges. The next boats *(MAS 430, MAS 432, MAS 433, MAS 434* and

MAS 437) were again derived from the SVAN 16m (52ft 3in) design, their 559·5kW (750bhp) Isotta Fraschini or Fiat engines resulting in a similar improvement in speed. A comparable armament was carried by the stepped Baglietto 16m (52ft 3in) *MAS 431* of 1931, but the Royal Italian Navy considered a number of experimental craft before deciding on a suitable type for mass production.

One such vessel was the 32m (104ft 6in) moto-torpediniera *Stefano Türr* launched in 1936. Big enough to carry three machine guns, four torpedo tubes, and twelve depth-charges, her semi-planing hull was built of Alclad (a Canadian aluminium alloy) which gave a full-load displacement of only 64 tonnes (63 tons) and a mere 0·6m (2ft) draught. It was hoped that her four 559·5kW (750bhp) Fiat diesels would produce a speed of 34 knots, but (probably because of the diesel power/weight problem) she never even reached 30 knots at sea, and therefore *Stefano Türr* saw no war service.

Meanwhile Isotta Fraschini's latest petrol engine was proving it was capable of 746kW (1000bhp). Some of the first ones were bought by Russia, but two were installed in each of the 17m (55ft 6 in) stepped boats being launched in Italy from 1936 onwards. Two 29·8kW (40bhp) Carraro auxiliary engines provided power for cruising and manoeuvring. These MAS-boats were usually known as the *Baglietto Velocissimo* type, although Picchiotti (River Arno), CRDA (Monfalcone) and other yards were involved as well as Baglietto at Varazze. Their overall dimensions were 17m × 4·3-4·6m × 1·3m (55ft 6in × 14-15ft × 4ft 3in), with a displacement of just over 20·3 tonnes (20 tons). They carried a 13·2mm machine gun and depth-charges, but their main armament comprised the usual two side-launching racks and clamps, each holding one 45cm (17·7in) A110/450 torpedo. This weapon was 5·3m (17ft 3in) long with a 110kg (242lb) warhead, and a range of 2000m (2180yd) at 38 knots. The series of boats ran from *MAS 501* to *MAS 525,* while a replacement for *MAS 424* was a near sister.

The next series, beginning with *MAS 526* was very similar, being 18·7m (61ft) long with a slightly bigger displacement. By September 1939 the series had reached *MAS 550,* the first MAS-boat with a metal hull. These boats could make over 40 knots in shallow coastal waters but they were not so effective in heavy weather. Only wartime conditions would show whether the torpedo dropping arrangements were satisfactory. Nevertheless, the Royal Italian Navy did have sixty-seven fast fighting boats available, based at Taranto, Naples, Messina, Augusta, Palermo, La Spezia, Sardinia, Tripoli, the Adriatic, the Dodecanese, and the Red Sea.

In direct contrast was Japan, who although another great naval power, had just two Thornycroft CMBs. One had been captured at Canton in 1938, and was a 16·8m (55ft) boat of the same class as *MTB 26* and *MTB 27* in the Royal Navy's flotilla at Hong Kong.

America, on the other hand, was showing an interest in fast fighting boats. The United States Navy had considered various designs during the First World War, but they were not followed up as it was unlikely that such craft would serve any purpose in American coastal waters so far from the war zone. Instead, seagoing submarine chasers were built, while motor launches were prefabricated or constructed for Britain, Russia, France, and Italy.

Several boatbuilding firms thus had experience of naval requirements, and motor boat races had stimulated development. Operational experience had been provided by bootleggers, who had become so proficient in employing old CMBs and other fast boats that the United States Coast Guard acquired a total of thirty-one destroyers to intercept them. Fifteen 23·8m and 24·4m (78ft and

80ft) motor cutters also served in the United States Coast Guard, their construction continuing until 1937.

Meanwhile the United States Navy had purchased two Thornycrofts of their own, the CMBs performing various experiments and other tasks until the 1930s. Aware of foreign MTB developments, the United States Navy gave further thought to their employment at the end of 1936. Of little immediate value, they could release other warships from coastal defence which might even be used offensively if such an unlikely situation ever arose. Indeed, within a few years, their advocates were claiming that 60-knot mosquito boats would be wrecking troop transports, battleships, and aircraft carriers with impunity.

In 1938 prizes were offered for the best craft 16·5 or 21·3m (54 or 70ft) in length. The yacht designers, Sparkman and Stephens, won the competition with a 21·3m (70ft) project for what became designated as a patrol torpedo boat (or PT boat). By the middle of 1939, *PT 1* and *PT 2* had been ordered from Miami Shipbuilding Co, and *PT 3* and *PT 4* from Fisher Boat Works of Detroit. All four were designed by Professor George Crouch, being 18m (59ft) long and displacing 25·4 tonnes (25 tons) for easy shipment overseas. The Miami pair were to be powered by two Vimalert 895·2kW (1200bhp) engines, while the other two would have two 895·2kW (1200bhp) 12-cylinder marine Packards being specially developed for MTBs and burning 100-octane petrol. Philadelphia Navy Yard undertook the construction of the 24·7m (81ft) *PT 7* and *PT 8*. Four 671·4kW (900bhp) Hall-Scott engines would power the government-designed *PT 7*, her aluminium sister having four 746kW (1000bhp) Allison V-motors arranged as two X-engines. They would be drag-started at 15 knots by a 410·3kW (550bhp) Hall-Scott, which was also used for going astern. *PT 5* and *PT 6* were based on the Sparkman and Stephens design. Both were 24·7m (81ft) long with three Packard engines developing a total of 2797·5kW (3750bhp) and 3021·3kW (4050bhp) respectively, for speeds of 35-40 knots.

These PT-boats of the 2nd Squadron were 21.3m (70ft) Elcos, based on the Scott-Paine *PT 9,* **except** *PT 8,* **which was a 24.7m (81ft) design from the Philadelphia Navy Yard. Here the boats are leaving the Washington Navy Yard in the winter of 1940-1.** *Imperial War Museum*

The contract for both was awarded to Higgins Industries of New Orleans. This firm was very much the personal creation of Andrew Jackson Higgins, who founded the Higgins Lumber and Export Company. He soon established a ship repair yard and then entered the boatbuilding business specialising in rugged work-boats for shallow-water operation. A semi-tunnel protected the propeller from damage by weed, snags, and shoals but aerated water sucked into the tunnel caused a loss of propeller efficiency, the sort of challenge which Andrew Jackson Higgins could never ignore. After a series of theoretical tests and practical modifications, *Eureka*, completed in 1937, embodied a horizontal head-log instead of the normal vertical stempost for greater strength forward, and a spoon bow to trap aerated water under the forefoot and reduce friction. As her V-shaped hull form ran aft, it changed to a reverse curve thus filling the semi-tunnel with solid water on which the propeller could bite. This groove also helped to keep the boat upright, when turning so that *Eureka* could back off a beach and turn in her own length between waves. This facility proved ideal for small landing craft when these became necessary but Higgins wondered whether *Eureka*'s speed and manoeuvrability would be equally useful in an MTB.

Hubert Scott-Paine from British Power Boat now re-appeared on the American scene. His 21·3m (70ft) private venture MTB had not been accepted by the Admiralty, but Henry R. Sutphen of Elco had learned of her capabilities. He travelled to England, decided the design was suitable, and bought the boat at his own expense. The United States Navy approved the purchase, reimbursed him, numbered the boat *PT 9,* and shipped her to New York where she arrived on 5 September1939.

Scott-Paine took another of his 21·3m (70ft) private venture boats to Canada where she performed successfully and was commissioned into the Royal Canadian Navy as *CMTB 1.* Plans were drawn up for the construction of a factory for the production of sister craft by Canadian Power Boat at Montreal.

This Canadian attention represents the appeal of MTBs to those countries which could not afford big ships and large crews and yet needed some maritime defence against aggression at a time of international crisis. A few, fast, elusive boats, each packing a big punch could easily deal with any hostile battleship or cruiser that dared violate their neutrality. The Rumanian and Chinese orders for Vosper 21·3m (70ft) and Thornycroft 16·8m (55ft) boats have already been mentioned. Rumania had earlier acquired several First World War Elco MLs, while back in 1921 China had purchased two old Orlando 12·2-tonne (12-ton) MAS-boats. Two more of this class (now named *Sisu II* and *Hurja*) were serving in the Finnish Navy, who hoped to buy two 16·8m (55ft) Thornycroft CMBs. The same firm was building another five of these craft for the Philippines, while the Royal Norwegian Navy had ordered Nos *1, 2, 3,* and *4* from British Power Boat, and *5, 6, 7,* and *8* from Vosper. The two types were respectively 19·2m (63ft) with two 820·6kW (1100bhp) Rolls-Royce Merlins, and 18·3m (60ft) with two Isotta Fraschini engines of 820·6kW (1100bhp) each. The Royal Swedish Navy had discarded *N 1* and *N 2* (ex-Orlando 12·2-tonne [12-ton] MAS-boats) in 1927. They now decided to buy new vessels, splitting their order between British Power Boat – *T 1* and *T 2* of 19·2m (63ft) – and Vosper – *T 3* and *T 4* of 18·3m (60ft). The latter pair were delivered first and armed in Sweden. Two of the same design were being built for Greece.

The Adriatic had seen many successful fast fighting boat operations during the First World War. Yugoslavia was aware of these boats' potential in this area and had acquired two 16·8m (55ft) Thornycroft CMBs in 1927.

Uskok and *Cetnik* were still in service in 1939, having been joined recently by *Kajmakcalan, Durmitor, Orjen, Velebit, Dinara, Triglav, Suvobor* and *Rudnik*. Among the earliest MTBs to be named, these latter 28m (91ft 9in) S-boats had been built by Lürssen but were powered by three 452·2kW (950bhp) Mercedes-Benz petrol engines, not diesels as in the German craft. It is interesting that names have usually been more readily allocated to minor warships in the world's smaller navies. The Spanish Nationalists renamed their two MAS-boats *Napoles* and *Sicilia* when they were transferred from Italy in 1937.

Soon after the First World War the Royal Netherlands Navy acquired four Thornycroft 16·9m (55ft) CMBs for service in the East Indies. *TM 3* had to be scrapped in 1933 after being damaged and an almost identical replacement was built at Sourabaya in 1938. Wooden craft were not entirely suitable in the tropics unless sheathed with copper which could be prohibitively heavy. So when the original *TM 1, TM 2,* and *TM 4* became due for scrapping, it was intended that their replacement series beginning with *TM 4* should have steel hulls. They would also be built at Sourabaya, being slightly bigger than their predecessors with three 335·7kW (450bhp) Lorraine Dietrich engines from Dornier Wal flying boats. A wooden MTB was judged perfectly adequate for Dutch home waters. Accordingly, the 21·3m (70ft) *TM 51* was ordered from British Power Boat to serve as a prototype for sisters being built by Werf Gusto at Schiedam. This latter firm would also construct eight German 28m (91ft 9in) S-boats.

Ireland was one neutral country that acquired MTBs to further her foreign policy, besides being part of her coastal defence force. They were as 21·3m (70ft) Vosper hard-chine design, with Isotta Fraschini engines, but were built by Thornycroft so that their overall length came out as 22m (72ft). If Ireland did become embroiled in the expected war, then her six MTBs would be a welcome addition to the forces of her ally.

So, as the Second World War began in September 1939, most navies had anticipated that the conflict would involve some sort of fast fighting boat warfare. Generally, though, the Great Powers were reluctant to devote much of their strength to such craft, hoping that experience would show which prototypes ought to be developed and produced in quantity, providing they had the time and industrial capacity to do so. For small neutrals, MTBs and MASBs seemed an inexpensive means of coastal defence, torpedoing battleships, and depth-charging submarines. Little thought seems to have been given to defence against attacks by these light coastal forces. A capital ship's secondary armament might be effective against MTBs, but the best strategic protection might be to keep out of coastal waters altogether. It does not seem to have been considered that fast fighting boats might engage each other, and nobody dared openly admit that sneak torpedo attacks might be made against merchantmen. Even to have hinted that the other side might do so could well have been regarded as a provocative breach of faith in an atmosphere charged with crisis and pervaded with naval agreements.

Yet when hostilities began, there was no dramatic encounter between MTBs and battleships, nor did MASBs race around a submerged U-boat pounding her to destruction. Because of mines, the threat of air attack, the need to be at some distant strategic point, and for many perfectly good reasons of their own, capital ships and submarines kept to deeper, rougher waters beyond the range of fast fighting boats.

Only in the Baltic was there the possibility of action, where the German torpedo boat force included the 1st S-Boat Flotilla with their depot ship *Tsingtau*. S-boat guns accounted for one small Polish vessel early on, but they found little employment in the shore bombardments that characterised the Polish campaign. They were soon handicapped by the winter ice of the Baltic, while their sister flotilla at Wilhelmshaven suffered from heavy weather in the southern North Sea.

To improve their seaworthiness and thus maintain a higher speed in all conditions, the next series of S-boats were built with raised forecastles. The torpedo-tubes were thus completely enclosed and hence screened from weather and enemy fire. This layout proved so advantageous that by the end of the war almost 200 near-sisters to *S 26* had been completed by Lürssen, by Schlichting (Travemunde) and by Danziger Waggonfabrik. Their overall length increased to 35m (115ft) and diesels of 1865kW (2500bhp) were eventually installed so that the S-boats always had plenty of power in reserve. They carried reload torpedoes and their gun armament was also augmented in response to combat requirements as the war progressed.

The following April, 1940, the S-boat depot ship *Carl Peters,* with *S 19, S 21, S 22, S 23* and *S 24,* took part in the Norwegian Campaign, being assigned to Group 3 bound for Bergen. *Tsingtau, S 7, S 8, S 17, S 30, S 31, S 32* and *S 33* went to Kristiansand and Arendal with Group 4. They were mainly employed as ship-to-shore ferries, but also patrolled fjords, a task which resulted in the destruction of the Norwegian torpedo boat *Sael.*

Little seemed to be happening on the Allied side. With Italy remaining neutral, it was decided that the 1st MTB Flotilla was merely redundant at Malta, and they came home via French canals and rivers. *MTB 6* foundered while under tow off Sardinia, but the others arrived at Portsmouth in December 1939. There Admiral Sir William James suggested to Winston Churchill (then First Lord of the Admiralty) that a piece of derelict Royal Air Force land be purchased as a Coastal Forces base. This was the future HMS *Hornet*, but for the present the 1st MTB Flotilla and their trawler depot ship *Vulcan* moved on to Felixstowe, where HMS *Beehive* was being established. For most of the bitter winter of 1939-40, the MTBs were employed on coastal patrols and submarine hunts, dashing out on air/sea rescue missions, or wallowing along as escort to some coastal convoy. This last seemed a monotonous waste of fuel and boats, but there were always more merchantmen than escorts available, so every small warship was pressed into service, especially those that did not seem to be doing anything else. These tasks, though providing invaluable seamanship, maintenance, and design experience, did little to develop the specific skills and theories of MTB attack. Fast fighting boats might have been profitably employed during the Norwegian campaign of 1940 but there was no way of establishing a Royal Navy base for them in time. This revived suggestions for an MTB carrier, a concept which has occasioned interest since the earliest days of torpedo boats.

The Admiralty used this period of the war to prepare the experimentals for service, complete the boats under construction, and requisition those on foreign order. Not all the buyers were entirely happy with this arrangement, particularly the Dutch, whose *TM 51* was undergoing acceptance trials off the British Power Boat yard at Hythe on Southampton Water. Lieutenant-Commander O. de Brooy RNethN was allowed to complete these trials so that the relevant data could be used in the construction of *TM 52* and *TM 53* at Schiedam. It so happened that *TM 51's* final high speed tests ended so far to the east that it was necessary for her to put into the nearest port, which happened to be Dutch; *TM 51* had joined the Royal Netherlands Navy. The requisitioned Vospers joined the 4th MTB Flotilla, while the Thornycroft CMBs went to the 10th MTB Flotilla, both being based at Felixstowe.

Some of the disappointed countries whose vessels had been requisitioned by the Admiralty obtained craft from other suppliers like Italy, who sold the 17m (55ft 6in) *MAS 506, MAS 508, MAS 511* and *MAS 524* to Sweden in February 1940. An 18·7m (61ft) boat of the MAS 526 series went to Japan, who also built a 19m (62ft 3in) CMB of her own based on the 16·8m (55ft) Thornycrofts she had acquired earlier.

America, too, entered the market, Higgins gaining a contract for twelve 21·1m (69ft 3in) Finnish patrol boats. Not requiring such a high performance, they would be equipped with three of the less powerful 373kW (500bhp) Hall-Scott engines, giving them a speed of 27 knots. Andrew J. Higgins had not felt content with the Sparkman and Stephens design and by 1940 Higgins Industries were working on a second *PT 6* embodying their own *Eureka* experience. The new boat proved most satisfactory, unlike the other experimentals. *PT 3* and *PT 4* seemed too small, while *PT 7* and *PT 8* were overweight, mainly because destroyer equipment had been installed. *PT 1* and *PT 2* were still awaiting their engines. Meanwhile Scott-Paine's *PT 9* had been so successful that detailed measurements were taken from her and she served as the prototype for twenty-two PT-boats and submarine chasers to be built by Elco.

9 THE DEMANDS OF WAR

Then, suddenly, there was a use for MTBs after all. The Germans were the first to exploit the opportunity, although this was as much due to the fortunes of war, as to deliberate policy. Just as any navy cooperates with the army, so the *Kriegsmarine* was expected to work with the *Wehrmacht,* covering its coastal flank during the attack in the west. The S-boat flotillas had practised offshore interception and they had a part to play in this campaign. But by the end of April 1940 so many larger German warships had been lost, damaged, or were still engaged in the battle for Norway that the S-boats assumed even greater significance in the forthcoming operation.

During their first westward patrol, the 2nd S-boat Flotilla under *Kapitänleutnant* Rudolph Petersen, encountered a force of British cruisers and destroyers in the southern North Sea on the night of 9/10 May 1940. Fire was exchanged, but *S 31* was able to carry out an unobserved and silent stalking attack from the dark side. *Kelly* was severely damaged by *S 31's* torpedo, while *S 33* came in so close that she actually collided with the crippled destroyer.

For the Royal Navy's MTBs, the campaign in France and the Low Countries meant an unending series of air attacks while trying to complete whatever hurried task had just been assigned to them. A temporary base was established at Ijmuiden for patrols preventing German flying boats from using the Zuider Zee. It was soon abandoned, the MTBs returning with refugees. Another vessel to leave Rotterdam was *TM 51,* which had defended the Maas Bridges so valiantly that her captain, Lieutenant J. van Staveren RNethN was awarded the Dutch *Militaire Willemoorde.* The rescue of crews from the Zeebrugge and Ostend blockships was another assignment carried out by British craft.

Then came the evacuation from Dunkirk, *MASB 6, MASB 7, MASB 10, MTB 16, MTB 22, MTB 67, MTB 68, MTB 102* and *MTB 107* serving as despatch boats, ferries, and rescue launches. They were also ordered to ward off E-boats, a term apparently standing for enemy war motor boats. Originally referring only to *Schnellboote,* it was soon applied to most hostile coastal forces, regardless of nationality and type, in the same way that U-boat applied to any suspicious submarine.

This warning about S-boats had been issued because they had moved their base round the coast into occupied territory and immediately took the offensive with mine and torpedo. *S 21* and *S 23* shared in the destruction of the French destroyer *Jaguar,* while on 28 May 1940 *S 34* sank the small coaster *Abukir.* That night the drifters *Nautilus* and *Comfort* were en route for Dunkirk when they encountered one or more S-boats. *Nautilus* opened fire with her Lewis gun and the S-boat disappeared in the darkness. It may well have been *S 30* for within a few minutes, she had fired two torpedoes at the destroyer *Wakeful*. Full of troops from Dunkirk, *Wakeful* was blown apart amidships and sank immediately. The two drifters recovered twenty-five men, but then the destroyer *Grafton* was torpedoed by a submarine. As she settled, a low shape was glimpsed in the darkness and fire was directed at what was assumed to be the attacking E-boat. The big minesweeping sloop *Lydd* also opened fire, ran down the sighting, and rammed what proved to be the drifter *Comfort*. There were few survivors from her.

That night was typical of the experiences of both sides when attacked by MTBs. Confusion predominated from the instant that roaring engines, foaming wakes, and explosions rent the night. The same boats would be sighted in several places around the convoy, and both sides consistently exaggerated the number of assailants. Friendly escorts were fired on and anything crossing one's bows was liable to be rammed. Explosions, gunfire, smoke, signals, wreckage, and cries from the water all contributed to the chaos. The crew who were least confused, kept a good lookout and positively identified and fired at an approaching MTB, stood the best chance of survival.

Any gunfire might spoil the enemy's night vision for the next thirty minutes, or cause the torpedoes to be launched a few yards too far away or a few degrees off target. Having fired her torpedoes, there was no point in staying, and the MTB turned away at high speed behind a smoke screen, her guns blazing. With flashes of flame all over her, trailing smoke, and leaping about among towering waterspouts, the hostile craft had obviously been mortally hit and would never get home. So both sides consistently exaggerated their claims of MTBs destroyed, which when published delighted the opposing propaganda ministries, but caused considerable problems for staff officers trying to ascertain the exact results of the night's work.

On 30/31 May 1940, *S 23* and *S 26* torpedoed the French destroyer *Sirocco,* her sister *Cyclone* having her bows blown off by *S 24*. Subsequently *S 34* accounted for the trawlers *Argyllshire* and *Stella Dorado,* but there were no encounters with the Royal Navy's Coastal Forces. For them the battle remained an air/sea one. Although they were often nearmissed by bombs, no fast fighting boats were lost during the evacuation (known as Operation Dynamo) a tribute to the toughness of their construction and a reassurance to those who may have wondered whether they were too fragile for war. Because of their manoeuvrability and inherent resilience to shocks, all fast fighting boats stood a good chance of surviving bombs in the open sea, but they could suffer heavy casualties from low-level machine gunning, and were extremely vulnerable in dock. Early air raids on Boulogne soon persuaded the German naval authorities that S-boats needed special protection when in harbour. The result was the construction of bomb-proof concrete pens which became a feature of S-boat and U-boat bases.

The Germans were the first to use MTBs regularly against merchantmen, principally because they had a ready target – the English coastal convoys from and to Newcastle-Upon-Tyne. Each collier carried about 2032 tonnes (2000 tons) of coal, some thirty vessels per day being required to keep London in solid fuel. This was the total equivalent of 6000 lorries or railway waggons. And of course, all the other East and South Coast cities, towns, villages, factories, dockyards, and camps were equally dependent upon the safe arrival of their supplies. In these early days, too, ocean ships could still be routed through the English Channel.

On 19 June 1940 *S 19* and *S 26* torpedoed and sank the 3152·6-tonne (3103-ton) *Roseburn* off Dungeness, the same area where two more vessels were destroyed by *S 19* and *S 36* five days later. Then on 4/5 July 1940 Convoy OA 178, already reeling from a day of air attacks, was intercepted by S-boats west of Portland. One merchant ship was sunk (by *S 19* again) and two others damaged by *S 20* and *S 26*. The attacks continued, the trawler *Cayton Wyke* and the French repatriation ship *Meknes* being among those sunk by 25 July 1940. On that date, the S-boats tried an afternoon sortie against Convoy CW8 which was in disarray after being divebombed in the Straits of Dover. They were met by the destroyers *Boreas* and *Brilliant,* accompanied by *MTB 70* and *MTB 69,* who had been picking up survivors from the convoy. Accurate fire from the destroyers forced the S-boats to retire behind a smoke screen, which the MTBs tried to penetrate, only to come within range of shore batteries as they did so. All four British warships had been strafed by fighters, suffering some damage, but now they were subjected to severe divebombing. *Boreas* and *Brilliant* were both hit, but the warships managed to survive. After their withdrawal, *S 19, S 20,* and *S 27* succeeded in sinking three ships from the convoy.

From now on aircraft of both sides prohibited the deployment of fleet destroyers close to the Straits of Dover and prevented any effective surface movement at all during daylight hours. Night-time traffic was restricted to specific routes by minefields. The attack on this shipping now devolved upon the MTBs of both sides. It was hoped that as they skimmed over the surface they would be immune from both contact and influence mines but in fact this was not so. Mines were the third most important cause of loss to high speed attack craft during the Second World War. Even if the mine did detonate safely in a boat's wake it could have a disastrous effect on the next astern.

Early engagements between fast fighting boats achieved little. On one occasion *MTB 70, MTB 69* and *MTB 72* were on separate patrol just before dawn, when Lieutenant J. B. King-Church RN sighted six S-boats near Cap Gris-Nez. Besides her torpedo tubes, *MTB 70*, one of the Greek 21·3m (70ft) Vospers had six Lewis guns but only two Isotta Fraschini engines to provide spares for other boats. Her top speed was therefore only 27½ knots, so that after engaging the S-boats she was unable to escape until the approaching daylight forced the enemy to turn back. Although under fire for twenty minutes *MTB 70* had only received one graze. Joined by *MTB 69,* the two craft were identified by *MTB 72* as one S-boat, who attacked but her two machine guns jammed at the instant of opening fire.

The action of 13 August 1940 was similarly inconclusive. Three of the 18·3m (60ft) British Power Boat MTBs encountered a group of German MLs (R-boats). One of the MTBs was damaged in an attempt to ram an R-boat. Before taking her in tow *MTB 14* had to house the torpedo launching gear at the stern, but then six trawlers presented themselves as ideal targets. The damaged MTB could not accelerate for inertial launching of her torpedoes, while *MTB 14* could not get her outrigger ready. The chance was lost, but another attack by this flotilla *(MTB 17, MTB 14* and *MTB 15)* was more successful on the night of 8/9 September 1940. They could not find the convoy reported by air reconnaissance and were coming out from the shore during an RAF raid on Ostend, when *MTB 17* and *MTB 15* sighted merchant ships at anchor. Torpedoes were launched singly and hits were reported. Three ships were later seen to be wrecked, although it was not clear whether this could be attributed to the MTBs or to the bombing.

It was now apparent that the Germans were themselves having to run coastal convoys to build up their invasion forces for Operation Sealion and to maintain the industry and commerce of their occupied countries. As in Britain, complete reliance on land communications would have resulted in unacceptable congestion as well as increasing the danger from air raids (and later sabotage). Inland waterways have always featured prominently in European transport while it was more economical for the Scandinavian and Baltic traffic to proceed as far as possible before transhipment. So coastal convoys were passed along the Continental side of the English Channel and North Sea. But they were the only major convoys the Germans had to consider, unlike the Royal Navy who had to protect trade routes all over the world. The German convoys were therefore compact, consisting of just three or four vessels guarded by ten or so escorts. Yet they still had S-boats to spare to carry out raids against the British side where two or three warships tried to shepherd columns of twenty or more ships along a narrow fairway between the minefields.

From 4 September 1940 onwards, the East Coast of England became known as 'E-Boat Alley', the collier crews relating all sorts of stories about the dreaded E-boats. It was even said that the Germans had once put a party ashore on some lonely spot on the English coast, captured a country policeman on his bicycle, and taken him back to Germany for interrogation. It was certainly believed that the E-boats actually secured to navigation buoys. In fact, they merely remained in the vicinity of a buoy, waiting for a convoy to pass – but the effect was the same. An efficacious ruse, employed by both sides, was to lurk inshore hidden by the gloom of the land at a time when most danger was expected from seawards. Having selected their target, as many as twelve S-boats at a time fired their torpedoes and surged through the convoy heading directly for the safety of the open sea and a course for base.

By such tactics, the S-boats accounted for thirty-five merchant ships sunk and seven damaged between June 1940 and June 1941, some boats hitting two ships with two torpedoes. The destroyer *Exmoor* was sunk on 25 February 1941 while yet more vessels fell victim to mines laid by S-boats. Even so they were not without their own setbacks as on the night of 7/8 January 1941, when they failed to achieve any success at all against a convoy in the Thames Estuary. Waiting for favourable circumstances and a break in their escort and minelaying duties often meant that a week or even a month might go by without any S-boat torpedo attack on coastal convoys. Of course this was not known to the merchantmen and their escorts, and E-Boat Alley retained its sinister reputation for the whole of the war.

During this period the deficiencies of the Royal Navy's fast Coastal Forces were becoming apparent and the first remedies attempted. A V-formation was most suitable for MTBs when cruising, for then the boats kept clear of each other's pressure wave. But it was no good in the unanticipated crises of battle, for then they fouled each other's course and gunfire. The MTBs were often ordered to search too small an area, while boats were not always able to signal their sightings so that other groups could intercept. A stern approach seemed best, for a poorer lookout might be kept in this direction. But the Vosper boats could only do 7-9 knots on their quiet auxiliary engines, taking so long to overtake that they were bound to be seen. Coming in on unsilenced main engines was quicker but announced their presence, while the torpedo angle from this direction was so narrow that they would probably miss. An 'angle on the bow' so that the torpedo arrived at 90 degrees to the target's length was best, but unless the MTB found herself in this position it was unlikely that she would have enough silent speed to reach it unobserved, and, of course, the

18·3m (60ft) British Power Boat MTBs had to be doing full speed before they could launch their torpedoes at all. Generally torpedoes were fired at too great a range (as much as 3700m [4000yd]) and singly, the importance of not wasting these expensive weapons having been instilled into every prewar naval officer: when disengaging, smoke should be used in such a way that it did not hamper other MTBs or afford protection to the enemy and when regrouping, there should not be fatal mistakes in identification.

Some of these shortcomings could only by corrected by experience which could then be passed on in training. In October 1940, HMS *St Christopher* was established at Fort William for the general training of Coastal Forces personnel. Engineering instruction in high-powered petrol engines was eventually provided at Portsmouth and Portland, but it was over a year before the working-up of a newly commissioned boat could be carried out at Portland by HMS *Bee*. Reliable radio equipment was needed, while battle signalling had to be simpler. Dim blue-shaded lamps were used for single-letter challenge and reply, verbal communications security being improved by referring to other boats by their commanding officers' nicknames, a practice also employed by some S-boat flotillas.

The operational difficulties resulting from the abrupt cessation of the supply of Isotta Fraschini engines cannot be over-emphasised. The Vosper MTBs had been designed around three of these engines and now they were not available. Emergency improvisations did not fit into the structural, trim, feed, lubricating, and cooling systems devised for the Isotta Fraschinis, the result being slow speed, or unreliability, or both. Before the war the dangers of relying on a foreign source had been dismissed because the Royal Navy could always buy Rolls-Royce Merlins or other aero-engines. Now they were all required for the RAF, their priority guaranteed by the Ministry of Aircraft Production.

However, Coastal Forces did have a surplus of MASBs, which had few U-boats to find in shallow waters and had no success in hunting them even when they had been reported in the area. Some were converted to ASRLs, but the others under construction might be of some use against S-boats. Accordingly their depth-charge armament was reduced and the number of machine guns increased, being further augmented by one of the coveted 20mm Oerlikons.

The 6th Motor Gun Boat (MGB) Flotilla was formed in December 1940, proceeding to HMS *Beehive* in March 1941. At first they cooperated with convoy escorts and destroyers, but the risk of mistake was too great and the continual slow speed uneconomical. Soon MGBs were allowed to cover the whole of the southern North Sea, groups of two or three locating suitable points to ambush S-boats proceeding to and from the East Coast convoy routes. But both types lacked enough punch to do each other mortal damage. Running battles could last an hour. There were experiments with creosote sprays. Smoothbore Holman Projectors used compressed air to lob grenades and flares from the British boats, while both sides dropped depth-charges under their opponents' bows. Yet damaged boats still got back to base. Heavier armament was needed but MGBs had a low priority and even forfeited their Oerlikons for a while. At the same time, continuous running when heavily loaded started to strain engines and hulls.

In August 1941 Lieutenant-Commander Robert Peverell Hichens RNVR was promoted to command the 6th MGB Flotilla. He was not the only MGB commander, nor even the only famous MGB commander, but this courteous gentleman and dedicated warrior was one of the inspirations of Coastal Forces.

A Cornish solicitor with small boat and motor-racing experience, he had won the Distinguished Service Cross at Dunkirk and was to win the Distinguished Service Order twice and the DSC twice more. He soon began to press for specially designed MGBs with a powerful armament. In the meantime the idea of steam gun boats (SGBs) had been suggested, quiet enough to stalk a convoy, yet with the high speed of an MTB and the guns necessary for dealing with S-boats.

By now the Royal Italian Navy was also employing a new and unorthodox type of high speed attack craft. The MAS-boats' first action had occurred on 14

A pair of early British Power Boat 21.3m (70ft) MGBs, built as Masbies in 1940 and converted a few months later. This 1942 picture gives an indication of the variations in armament and construction details. Ahead lie, to starboard, another British Power Boat MGB and to port a British Power Boat ASRL.
Imperial War Museum

June 1940, when French cruisers and destroyers had bombarded Vado and Genoa at daybreak. The 13th MAS-boat Flotilla, backed up by the torpedo boat *Calatafimi*, had come out from Genoa but without success. Similar disappointment attended *MAS 536* and *MAS 537* when they attacked two British cruisers and two destroyers in the Kaso Strait near Crete at dawn on 4 September 1940; *MAS 537* was sunk by *Ilex*.

Now the *barchini esplosivi* (explosive little boats) entered service. Officially designated *motoscafi di turismo* (tourist motor boats), the MT-boats were 5·6m (18ft 5in) long, with a displacement of 1·1 tonnes (1·1 tons). A 70·9kW (95bhp) Alfa-Romeo outboard drive engine gave a speed of 33 knots over a range of 80 miles. Built by Baglietto and CABI-Cattaneo of Milan, they formed the Surface Division of the X Flotilla under, respectively, *Capitano di corvetta* (Lieutenant-Commander) Giorgio Giobbe and *Capitano de fregata* (Commander) Vittorio Moccagatta. The MT-boats were especially intended for use against shipping moored within a defended harbour. They were tiny targets and it was hoped their high speed would send them skimming over mines and under shore batteries' maximum depression. Their outboard drives were hinged so that they bounced across nets, booms, and obstructions. Within 100m (100yd) of his objective, the pilot armed a 330kg (726lb) charge of high explosive, locked the rudder, and threw himself out of the boat on his backrest. The *barchino* hit and either blew up immediately or was split in half by a ring of small charges around the hull; the warhead sank and was detonated by a hydrostatic fuse. If he was lucky, the pilot had about five seconds to clamber on to his backrest before the explosion's shock-wave crushed his body. If he was very lucky, he might then escape the general crossfire and become a prisoner of war.

On 25/26 March 1941 the destroyers *Francesco Crispi* and *Quintino Sella* sailed from Leros to launch six MT-boats against British shipping in Suda Bay. The tanker *Pericles* was disabled and a hole blown in the 8382-tonne (8250-ton) cruiser *York,* who settled on the bottom with her superstructure still above water. Salvage proved impossible and was abandoned after further damage had been inflicted by German air raids on Crete. These later strikes also wrecked five 16·8m (55ft) Thornycrofts, the major part of British Coastal Forces in the Mediterranean.

Twelve nights later (7/8 April 1941) and 1700 miles away, *MAS 213,* accompanied by *MAS 216* (both built in 1918), put a torpedo into the cruiser *Capetown* from a range of 300m (300yd) during the British attack on Massawa. The Red Sea boats were then scuttled, but the Italian Navy in the Mediterranean was soon receiving the surrender of six Yugoslavian S-boats originally built in Germany. They were eventually classed as *motosiluranti* (motor torpedo boats) and formed the pattern for a series of thirty-six CRDA boats constructed at Monfalcone. Instead of diesels, they were given three Isotta Fraschini petrol engines of 820·6kW (1100bhp) each, resulting in a speed of 33-34 knots. Some later ones carried two 45cm (17·7in) torpedoes aft as well as two 53·3cm (21in) torpedo-tubes on the forecastle, but they were otherwise identical to the German 28m (91ft 9in) S-boats.

On the night of 23/24 July 1941 *MAS 532* and *MAS 533* were sent from Pantelleria against the Substance Convoy from Gibraltar to Malta. Radar gave some warning of the MAS-boats' approach and so did hydrophones, but the escorting destroyers' starshell lacked flashless propellant. Their gunnery control personnel lost their night vision every time it was fired, while searchlights illuminated the convoy as much as the attackers. Still the Tribal destroyers in the escort had come from Plymouth where single 2-pounders had been mounted in the bridge wings for rapid-fire defence against E-boats. Several hits were reported on the MAS-boats and *Sydney Star* was the only vessel damaged by their attack.

Two nights later the *barchini* staged their next operation. A 32-knot sloop would carry nine of them and a *motoscafo di turismo siluranti* (tourist torpedo motor boat). This MTS-boat was larger than a *barchino esplosivo,* being 7·2m (23ft 6in) long with a displacement of 1·8 tonnes (1·75 tons). She had one Alfa Romeo 67kW (90bhp) outboard drive engine, giving a speed of 28 knots over 90 miles. These craft performed a similar function to the MT-boats, but were not expendable. They needed a crew of two to handle two 45cm (17·7in) torpedoes in telescopic tubes aft, plus smoke-making gear. *Diana* would be accompanied by *MAS 451* and *MAS 452*, the latter towing a small electric-engined boat, herself transporting two human torpedoes. When 20 miles from Malta, *Diana* would launch the *barchini* to be led on by the MTS-boat. *MAS 452* would slip the electric boat five miles out. The *barchini* would be on their own 1000m (1000yd) from Valletta, while one human torpedo would head for the submarine base near Sliema. The other would blow a hole in the net defence suspended from a two-span bridge over a narrow side entrance into Grand Harbour. If he failed, two MT-boats would do it. Through the gap would go the other *barchini*, their charges set at various depths. The whole operation would be covered by an air raid and would stand a good chance of success, for MAS-boat reconnaissance had shown that small craft could approach unchallenged.

On the night of 25/26 July 1941 the force went into action with few of the initial setbacks which so often bedevil such complicated schemes. A defect did develop in the Sliema-bound human torpedo which cleared the area, her pilot being captured next day. However it was not known that *Diana* had appeared on and then disappeared from the Malta surface radar screens, indicating that something suspicious was going on. Nothing else had been registered, but the defences were ready.

The first explosion came at 4.25am, but the human torpedo failed to penetrate the net. The first MT-boat did not explode, and the next *barchino* pilot rode his boat in and blew up with her, the detonation bringing down one of the bridge spans and completely blocking the channel. Under very heavy fire, three boats were scuttled, one with her pilot still on board, three were sunk and the pilot of the other was knocked overboard, his boat roaring off to be captured later (and subsequently preserved in Malta's War Museum). It was now daylight, three squadrons of Hawker Hurricanes scrambling to take part in the battle. Outnumbering ten Macchi fighters, they strafed *MAS 451* and *MAS 452,* who were towing the two small boats. *MAS 451* blew up, leaving nine survivors who were later captured. *MAS 452* was disabled, and *Capitano di fregatta* Moccagatta, *Capitano di corvetta* Giobbe (both awarded a posthumous Gold Medal for Valour), and six others were killed. Eleven men returned to *Diana* on board the MTS-boat. *MAS 452* remained afloat for some time, providing temporary refuge for a Hurricane pilot who had been shot down.

Meanwhile the development of fast attack craft was proceeding in the United States of America. Some of the experimental PT-boats had been commissioned as MTB Squadron 1 in July 1940. By the winter of 1940, Elco had completed the 21·3m (70ft) boats numbered *PT 10* to *PT 19.* Others were intended as high-speed submarine chasers, being known as *PTC 1* to *PTC 12.* Elco were now working on a larger but similar 23·5m (77ft) series to carry 53·3cm (21in) torpedo tubes. The firm of Huckins had brought out their 22m (72ft) *PT 69,* while another competitor was Higgins' *PT 70,* of 23·2m (76ft).

With all these rival designs to choose from, the United States Navy organised a series of trials culminating in a 190-mile race off the New England coast.

Two Elco boats came first and second in the Plywood Derby, followed by the Higgins, and then the Huckins boats. The best average speed was 39·7 knots, but most boats suffered some structural damage due to the bad placing of ballast as simulated armament. On a re-run in waves of 2-5m (6-15ft), an Elco boat again won at 27·5 knots, but there was not much in it. Certainly the Higgins boat had a smaller turning circle and could reverse course in 22 seconds, but she also had a tendency to dig her nose in. The Elcos were the driest, although they pounded the heaviest. It was found that the high-speed submarine chasers were no good at all. Underway, noise drowned the echoes; hove-to, the PTCs rolled so sharply they could not receive the echoes.

Contracts were awarded to Elco, Higgins, and Huckins for 40-knot hard-chine MTBs, powered by three silenced 12-cylinder Packards, 11355·9 litres

PT 103, **of the 5th Squadron, was the first Elco 24.4m (80ft) design, and was armed with an Oerlikon, a Thomson submachine gun, two Springfield rifles, thirteen Colt automatics and four 53.3cm (21in) torpedoes. Here she closes a seaplane tender** *(Pocomoke)* **to refuel during a trading run from Taboga (Panama) to the Galapagos Islands in November 1942.** *Imperial War Museum*

(3000 gallons) of 100-octane petrol giving them a cruising range of 500 miles. The 23·8m (78ft) Huckins series beginning with *PT 95* were used for training, while the Elco boats, starting with *PT 103,* would be of 24·4m (80ft) length. The Higgins 23·8m (78ft), 24·7m (81ft) overall, and commencing at *PT 71,* would carry the same armament as the others: four 53·3cm (21in) torpedo tubes, and 20mm guns. Anticipating wartime expansion, Andrew J. Higgins had already prepared factories for landing craft; now he embarked upon the assembly-line mass-production of PT-boats, mobilising the efforts of trade unions, civil rights, and scientists. Visits were arranged for American and foreign politicians and businessmen, while courses were provided for serving officers and men. Savings schemes and blood donors were encouraged, in addition to mottoes and posters ('The guy who relaxes is helping the Axis.') continually exhorting the Higgin plants to greater efforts.

Many of the early boats built by American firms, including those for Finland and the original British Power Boat *PT 9,* were transferred to the Royal Navy under Lend-Lease. Also transferred were Packard engines; at last British fast attack craft would have a good substitute for the missing Isotta Fraschinis. Or so it was hoped; but the earliest arrivals were still not reliable enough, suffering from a weakness in the crankshaft bearing casting. They also needed a greater oil flow, in addition to the perfect lubrication required by the V-drive and thrust blocks. Further loss of power resulted if 87-octane petrol had to be used instead of the 100-octane fuel they normally consumed. Then there were the usual problems involved in installing engines into boats designed for a different power-plant. These difficulties were temporarily exacerbated by air raids and the policy of dispersal. So, many British boats still could not do more than 30 knots and the North Sea flotillas could not always get to their patrol area and back under cover of darkness. Only the Dover MTBs could be sure of doing that. Even so, the number of engagements began to increase as the autumn of 1941 came on.

The S-boats, too, stepped up their operations as the nights became longer. Not that they had entirely disappeared from the English Channel and North Sea during the summer of 1941, but priority had been given to their role in Operation Barbarossa. Their opponents, the Russians, had employed American-built submarine chasers in the Black Sea during the First World War, and had also been on the receiving end of the Royal Navy's CMB raids into Kronstadt in 1919. Accordingly, they had not ignored fast attack craft, the aircraft designer A. N. Tupolev being one of the people taking an interest in their development. They were of three main types: Sch.4 was 18·1m (59ft) long overall, with two 484·9kW (650bhp) Wright T engines, and was based on Italian MAS-boats. The G 5-class was similar in appearance to a Thornycroft stepped CMB with two stern-launched 53·3cm (21in) torpedoes. The aluminium hull was 19·1m (62ft 5in) overall and was powered by two GAM.34 engines of 503·6-746·0kW (675-1000bhp) each. This class could suffer from considerable corrosion and vibration. The D 3-class was a continuation of the MAS design, carrying two 53·3cm (21in) torpedoes, one 20mm gun, and two 12·7mm machine guns. The D.3 type had three GAM.34 932·5kW (1250bhp) engines and was 21·6m (70ft 7½in) long. At the time of the German attack, the Russian Navy had forty-eight MTBs (known as torpedo cutters) in the Baltic, eighty-four in the Black Sea, and two in Arctic waters.

The majority of German S-boats were moved to Baltic bases and all were out on the night of 21/22 June 1941. The 1st and some of the 2nd S-Boat Flotillas helped escort minelaying squadrons into the Gulf of Finland, while the rest of the 2nd, the 3rd, and the 5th S-Boat Flotillas laid influence mines in the approaches to Soelo Sound, Moon Sound, Libau, Windau, and Irben Strait. Russian forces were sighted but the mines were laid on time and the S-boats could take the offensive as soon as hostilities commenced.

At 3.00am on 22 June 1941 the 3126·2-tonne (3077-ton) Latvian steamer *Gaisma* was torpedoed and sunk by *S 59* and *S 60,* the latter commanded by Lieutenant Siegfried Wuppermann, who had served in the Dunkirk operations. This success was paralleled by the 1st S-Boat Flotilla, who captured the 1199·9-tonne (1181-ton) *Estonia.* During the next weeks, the S-boats ranged ahead of the flank of the advancing Wehrmacht, attacking evacuation convoys and minelaying operations, while escorting German minelayers and often laying ground influence mines and explosive floats themselves. The Soviet torpedo cutters were similarly employed in cooperating with their land forces. The patrol boat *MO 238* was sunk by *S 44* off Hango on 22-23 June 1941, while near Steinort on 23 June 1941, *S 35*'s torpedoes missed the Russian submarine *S 3.* The S-boat followed up her attack with depth-charges and grenades, eventually sinking the submarine. *S 27* and *S 60* failed in an attempt on a Libau evacuation convoy and so did the 3rd S-Boat Flotilla when they encountered the submarine *S 7* in the Irben Strait two days later. They achieved better success when they hit the Estonian merchantman *Lidaza,* and fired a spread of torpedoes into Windau Harbour. Meanwhile *S 43* and *S 106* were blown up on Russian mines near Moon Sound.

They were avenged on the night of 26/27 June 1941 when the 2nd S-Boat Flotilla captured *TKA 47* off Backofen, one of the last vessels to leave Libau. *TKA 37, TKA 57, TKA 67* and *TKA 17* had all been able to make Dünamunde, although the other torpedo cutter *TKA 27* had had to be abandoned at Libau. That same night the 3rd S-Boat Flotilla was mining the Irben Strait when they encountered a Soviet force on a similar mission. The destroyer *Storozhevoi* was damaged and hits were reported on the submarine *S 10.* The S-boats which had fired their torpedoes, reloaded, and joined in the next attack, sinking the minesweeper *Shkiv.* Sorties by the 1st S-Boat Flotilla against Russian convoys off Eckholm in the Gulf of Finland failed, although the combined efforts of Soviet aircraft and torpedo cutters caused some damage to a German amphibious force entering the Gulf of Riga. The S-boats were also being deployed in these shallow waters, and on 15 July 1941 *S 54* with *S 47* and *S 58* with *S 57* tried to overwhelm a Soviet destroyer, but all their torpedoes missed. Nor did the 3rd S-Boat Flotilla do better in a sweep off Ösel five days later. But at last on 26-27 July 1941 *S 55* and *S 54* intercepted the destroyer *Smiely* in the Gulf of Riga and sank her on their second approach. Their flotilla-mates failed to torpedo two minesweepers anchored off Ösel, but then encountered the ice-breaker *Lachplesis* and destroyed her with gunfire.

The Russian torpedo cutters were mainly involved in evacuations, but they did try to take the offensive, although two sorties against German minesweepers were driven off with the loss of *TKA 122.* The Soviet destroyer *Artem* held her own against two S-boats on 2 August 1941 but the minesweeper *Pirmunas* was sunk at anchor by *S 58* later in the month. Then came an engagement between four torpedo cutters and German coastal forces, in which two of the latter were damaged. But little could be done to stem the German advance and soon fifteen torpedo cutters were assisting in the evacuation of Tallinn. All types of surface warships, auxiliaries and submarines took part. Most casualties were the result of air attack and the encircling minefields, many of which had been laid by S-boats. One merchantman was sunk by the Finnish MTB *Syoksy* who followed up this success by torpedoing the minesweeping trawler *Kirov.*

While the big ships of the Red Navy prepared for the defence of Kronstadt and Leningrad, the torpedo cutters tried to harass the German occupation of Ösel and Dagö. Others laid mines and evacuated the Hangö garrison to Suusaari. Up in the Arctic *TKA 11, TKA 12, TKA 13, TKA 14,* and *TKA 15* began attacking the Germans' supply between Kirkenes and Petsamo.

By the autumn and winter of 1941-2, the ice-free parts of the Baltic were completely in German hands, and they could redeploy their S-boat flotillas. They immediately made their presence felt back in E-Boat Alley, and by laying mines off Malta.

High speed attack craft had also been seeing service in the Black Sea where ten torpedo cutters had helped cover the evacuation of the Danube area on 18-19 July 1941. In August 1941 the 2nd Torpedo Cutter Brigade (totalling forty boats) was established at Odessa as part of the Red Army's coastal support. Among the opposing vessels were *Viscolul, Vigelia,* and *Viforul,* the Vosper 21·3m (70ft) MTBs supplied to Rumania in 1939-40. They failed to cut the Russian route from Sevastopol to Odessa, and in November 1941 *Viforul* and *Vigelia* were sunk on a minefield.

On 7 December 1941 the 0·5in machine guns on board *PT 23,* a 23·5m (77ft) Elco, accounted for two of the aircraft shot down at Pearl Harbor. Three days later *PT 31, PT 32, PT 33, PT 34, PT 35* and *PT 41* were manoeuvring at high speed to avoid Japanese bombs in Manila Bay. Within a week, British MTBs were engaging Japanese vessels at Hong Kong. During the eighteen months from the summer of 1940 to the end of 1941, fast fighting boats had proved they were capable of service in any part of the world and were able to perform any coastal task assigned them. The questions now were 'What were they best at?' and 'How could they do it?'

Russian MTBs of the G 5 class in the Black Sea, showing the strong influence of Thornycroft CMBs on Russian design.
Imperial War Museum

10 THE EXPERIMENTERS

If neutral nations and isolated colonies had believed that small cheap warships packing a big punch would deter any aggressor, they had been sadly disabused by the events of 1940-1. Although handled with valour, no high speed attack craft had significantly delayed the occupation of her country. Most had used their speed to survive during retreat. Admittedly the invaders had usually come overland, even though the last offshore islands might have had to be taken by amphibious assault. This was the case at Hong Kong, where the 2nd MTB Flotilla engaged Japanese minesweepers, landing craft, and sampans. Under heavy fire from ships, shore, and aircraft, the eight MTBs sank some of their targets and broke up one or two of the assaults. But they could not be everywhere at once, there was no time for repairs, and soon escape or scuttling remained the only alternatives.

The same was true of the Philippines. On the night of 18/19 January 1942 Lieutenant-Commander John D. Bulkeley in *PT 34* crept into Subic Bay at 8 knots and hit a Japanese transport with one torpedo at a distance of 450m (500yd). Actually, two torpedoes were launched but one stuck half out of its tube with its engine racing. Without surrounding water to cool it and retard the propellers, a torpedo on a 'hot run' can only go faster and faster until it disintegrates within a few seconds. In that short time, and in the dark, Chief Torpedoman Martino located the small valve in the side of the torpedo and closed the compressed-air line. The torpedo stopped, but by now waves were washing over its nose, turning the impeller vanes which armed the warhead. While *PT 34* hurtled and bounced out of Subic Bay, Chief Torpedoman Martino leaned out and jammed the vanes with a wad of paper. Eventually the torpedo fell into the sea.

They should have been accompanied by *PT 31,* but the squadron's lubricating oil had been sabotaged with sand and their petrol with wax deposits. Although the carburettors were cleaned hourly, they could still get choked and this happened to *PT 31.* The trouble was cleared up, then the cooling system broke down and the boat drifted on to a reef where she came under fire and had to be abandoned. On 8/9 April 1942, *PT 41* and *PT 34* successfully stalked the cruiser *Kuma* in the Tamon Channel between Cebu and Negros. Unfortunately, the one torpedo that hit failed to explode. Such setbacks were common when operating from improvised bases with exhausted equipment that could not be replaced from bombed-out stores. Even so, General MacArthur who, together with his family and staff, had been taken from Corregidor to Cagayan on Mindanao by the PT-boats, had been much impressed by their ability to survive and hunt in narrow, island-strewn waters. He advocated the expansion of the PT programme and arranged for Lieutenant-Commander Bulkeley (later awarded the Congressional Medal of Honor) and other personnel to be flown to safety so that their experience would not be lost.

Melville (Rhode Island) had been established as the MTB Squadrons Training Center, but during its life it also acted as the Repair Training Unit, a Command Course School, and a central drafting station for all personnel. Shakedown (or working-up) was undertaken at Miami and Taboga near Panama. Early war training was often conducted from tenders, the first being the converted yacht *Niagara*. Later PT-tenders could hoist damaged boats right out of the water either by crane or on an A-frame lift alongside. They proved invaluable in the remote areas of the Pacific.

One prewar claim for MTBs had been that they would deny coastal waters to capital ships. At 11.55am on 12 February 1942 Lieutenant-Commander E. N. Pumphrey RN set out from Dover to do this with *MTB 221, MTB 219, MTB 45, MTB 44* and *MTB 48*. Within three-quarters of an hour, they were intercepting the battlecruisers *Scharnhorst* and *Gneisenau*, the heavy cruiser *Prinz Eugen*, and twenty destroyers and torpedo boats. The 2nd, 4th and 6th S-boat Flotillas also took turns in supplying ten boats for a defensive screen on the port bow of the force. By laying smoke and by holding rigidly to their course, they prevented the MTBs from approaching closer than 4000m (4000yd) on the most suitable bearing. British and German boats exchanged gunfire at 200m (200yd) range. There was little damage for the sea was worsening, but it seemed certain that the vulnerable tubes would be hit before the torpedoes could be launched. Accordingly the MTBs fired their torpedoes, which all missed although the capital ships appeared to turn away to avoid them. The MTBs plus *MGB 43* and *MGB 41* were briefly engaged by the destroyer *Friedrich Ihn* who soon withdrew into the smoke to resume her place in the screen. Three MTBs from Ramsgate under Lieutenant D. J. Long fared no better.

Why did these boats fail, apart from the general circumstances surrounding the whole operation? Firstly, the MTBs assembled for stopping the break-out had been stood down. Then there was not enough warning for the whole available flotilla to be ready after the night's torpedo exercises and to reach the best position. Nor did they have much information about the composition of the enemy force. The boats themselves were either underpowered, beset by engine failure, or otherwise lacked the speed to evade the S-boats. Perhaps they could have got through if the S-boats had been kept off by a powerful force of MGBs or if their movements had been coordinated with an air strike. Most probably no daylight torpedo attack could have succeeded against such a determined, powerful, and fast-moving force. The chance of sinking an enemy capital ship – the ambition of every torpedo officer – had gone.

Such opportunities were still being presented to Axis boats in the Mediterranean; and they took full advantage of them. On the night of 14/15 June 1942, *S 54, S 55, S 56, S 58, S 59* and *S 60* were driven off on their first attack on the Vigorous Convoy from Alexandria to Malta. They carried out another stealthy approach, *S 56* firing her two torpedoes. The cruiser *Newcastle* evaded one but was hit by the other. She was able to keep going, but the destroyer *Hasty* was so badly damaged by *S 55* that she had to be sunk by her sister-ship *Hotspur*. Even more eventful were the hours from 3.00am to 5.08am on 13 August 1942. After two days of air and submarine attack, the surviving ships of the Pedestal Convoy from Gibraltar to Malta were passing close to Cape Bon. *MAS 553,*

MAS 556, MAS 560, and *MAS 562* – 18·7m (61ft) boats built since the beginning of the war – were picked up on radar and came under fire from four destroyers and two cruisers at the head of the convoy. The warships combed the torpedo tracks and all missed.

An attack by *S 59* (sent from Crete) also failed. *MS 16* and *MS 22* (two of the latest Italian craft and not yet fully worked-up) were lurking close inshore, partially concealed by the old wreck of the destroyer *Havock*. *MS 16*'s starboard torpedo ran wild after being launched at a distance of 800m (800yd) from the cruiser *Kenya*. The boats pressed in closer and fired their torpedoes at *Manchester*. Hit in the after engine-room, and with the after boiler-room and other compartments flooded, the cruiser stopped dead in the water. *MS 26* avoided *Dorset*'s ramming attempt, and her torpedo blew a hole in the merchant ship *Rochester Castle*. *MS 26* then made a second approach, but this time was lit up by starshell and searchlights from the destroyer *Pathfinder*. *MS 26* fired her remaining torpedoes and managed to escape under heavy fire. Meanwhile *MS 31* had torpedoed *Glenorchy* and was shadowing the convoy. *MS 25* and *MS 23* were held off by the escorts but *MAS 552* and *MAS 554* torpedoed and sank the cargo-ship *Wairangi*, while *S 30* and *S 36* disabled the American *Almeria Lykes*. *Santa Elisa*'s Armed Guard and their Oerlikons saved her from *MAS 557*, but she was hit by a torpedo from *MAS 564* and her cargo of petrol blew up. The last attack was made by *MAS 553* without success. *Glenorchy, Wairangi* and *Almeria Lykes* were eventually abandoned while *Manchester* could not be moved and had to be scuttled. None of the Italian and German boats had been sunk, although many had been damaged. Some remained in the area, rescuing survivors and hoping to find other targets. Only *MAS 556* encountered a group of warships returning from Malta at 20 knots during the night of 13/14 August 1942. Torpedoes were fired at *Kenya*, but they missed.

These successes seem to contrast sharply with the efforts of British MTBs against *Scharnhorst* and *Gneisenau*, but several points should be borne in mind. The Mediterranean attacks were ones of stealth, high speed being used to disengage and take up a fresh position. By the time they were in action, the MAS-boats and S-boats had some idea of the opposition they could expect. Their targets were really the merchantmen, and they could be soon driven off by the fire control of the warships. The attacks on *Newcastle* and *Manchester* were indeed well executed, but even these did not sink the cruisers outright. Finally, all these warships were tied to convoys, expecting or recovering from other forms of assault. The MAS-boats did not achieve anything against Force X returning from Malta unencumbered by any convoy.

The same was true of operations in the Black Sea in 1942. From 16 November 1941 the whole of the Crimea was in German hands except for Sevastopol. The nearest Russian bases were on the eastern shores of the Black Sea and it became routine for Soviet warships to rush in supplies and reinforcements, and return with wounded, shelling German positions on both passages. Occasionally transports, submarines, and the battleship *Parizhskaya Kommuna* took part, but the run was usually carried out by two or three cruisers and destroyers.

Icing and storms severe enough to damage the 8940·8-tonne (8800-ton) cruiser *Molotov* practically prohibited the winter employment of high speed attack craft, but the 1st S-Boat Flotilla and the 4th MAS-Boat Flotilla were transferred to Yalta by June 1942. Their operations against the Russian supply ships coincided with a massive land assault on Sevastopol which captured the city on 30 June 1942. The boats undertook attacks on 10, 12, 13, 15, 16, 19, 23, 24 and 25 June 1942, but they were either driven off by concentrated fire or failed to intercept the speeding warships. *MAS 571* torpedoed and sank the submarine *Shch 214,* while *S 102* accounted for a transport. That was all, although heavy losses were inflicted by German aircraft. The boats had better luck on 2/3 August 1942, when *MAS 568* and *MAS 573* caught *Molotov* and the flotilla leader *Kharkov* during a bombardment in Feodosiya Bay. A torpedo blew off 18m (60ft) of *Molotov*'s hull but she was able to make Poti, where she was repaired with a new section from the cruiser *Frunze.* Then came the evacuation from Novorossiysk, but again the S-boats and MAS-boats scored little success against warships, only three merchantmen being sunk by *S 102* and *S 28.*

These two, together with *S 27* and *S 72,* spent three nights early in September 1942 harassing Russian coastal forces trying to evacuate the Taman Peninsula. The S-boats claimed to have sunk nineteen small craft, although *S 27* was herself destroyed by her own torpedo. But against bigger warships, success still eluded them, as on the night of 22/23 October 1942 when four S-boats intercepted two cruisers, a flotilla leader, and three destroyers entering Tuapse. All their torpedoes missed and exploded ashore. Nor did the 1st S-Boat Flotilla hit any of the Soviet warships returning from a bombardment of Yalta and Feodosiya on the night of 19/20 December 1942. The two German and Italian flotillas resumed their campaign against Soviet coastal traffic in the spring of 1943. Barges and pontoons were sunk although their escort of patrol cutters was being strengthened. These had originally been designed in 1938 as submarine chasers, the MO 4-Class being 26·4m (86ft 6in) long with a displacement of 56·9 tonnes (56 tons). Three petrol engines produced a total of 1939·6kW (2600bhp), giving a speed of 25 knots. They had an armament of two 45mm and two 13mm guns plus depth-charges. These patrol cutters were intended to engage S-boats while torpedo cutters also tried to take the offensive. On one of their sorties they sank a ferry, and on another they fired torpedoes into Anapa on the night of 12/13 May 1943. Three S-boats were unable to prevent a repeat bombardment next night by the flotilla leader *Kharkov* and the destroyer *Boiky.*

Meanwhile American PT-boats were experiencing similar frustrations and successes in the Solomons. *PT 46, PT 48, PT 60* and *PT 38* had arrived at Tulagi on 12 October 1942, forming MTB Squadron 3 under Lieutenant-Commander Alan R. Montgomery USN. They went into action on the night of 13/14 October 1942 trying to attack a Japanese force of two battlecruisers, one cruiser, and seven destroyers bombarding Henderson Field. Once they had deployed, the boats lost touch with each other and could only locate the enemy warships' gun flashes and searchlights. At 10 knots, *PT 38* fired four torpedoes from 400-200m (400-200yd) and saw one hit on a cruiser. *PT 60* launched two torpedoes and claimed that both hit, but she was under heavy fire from destroyers and eventually ran aground to be towed off next day. *PT 46* and *PT 48* narrowly escaped collision with each other and with one of *PT 60*'s destroyers. Then the shore bombardment ceased and no further targets could be found. The Japanese had not suffered serious damage, but they did claim to have sunk fourteen out of nineteen PT-boats.

When possible the growing number of PT-boats continued their efforts against the 'Tokyo Express' during the following months, but in common with other forces in the area, so much conspired against them. The early bases were primitive, the boats having to be refuelled by hand, straining the 100-octane petrol through chamois leathers to remove the water. All electrical circuits had to be off during this lengthy process which was frequently interrupted by air raids. When completed, the boats were hidden under bushes or mangrove trees. This meant that rats, cockroaches, scorpions, lizards, flies, crabs and spiders came on board. A continual battle was fought against grime and mould, while bedding and clothes rotted in the humid heat.

Although some were able to fix up accommodation ashore in abandoned huts, officers and men usually lived in the cramped conditions on board. There were no big refrigerators, so they supplemented a monotonous diet by scrounging and fishing. The drinking water was heavily chlorinated, but even so dysentery was common and everyone suffered from skin troubles, and became tired and irritable, seemingly in a haze of stupor which sometimes led to quite silly mistakes. There was rarely the excitement of combat as for most of the time the PT-boats were employed on carrying mail, supplies and reinforcements for the Army or on the transfer of equipment and personnel between bigger ships. With little time for slipping and lengthy maintenance, marine growth soon cut speed. Sometimes only one engine out of three was working, while the strain of last-minute reversal occasionally proved too much, so that a boat carried straight ahead when berthing, with resultant damage to herself and the flimsy jetties. Then too, sudden rough weather could mean broken limbs, facial injuries, and gear adrift.

When on patrol, the PT-boats usually cruised on one engine to reduce noise and wake. The two shut down were turned over from time to time to

A Higgins PT-boat on the night exercises organised by Commander Allan R. Montgomery. *Imperial War Museum*

make sure they were working and to take the boat back to her designated patrol position. Although they might be told if a Japanese force was expected, the exact route through the islands could not be forecast, and most patrols were only enlivened by rain showers and the dark shapes of rocks and islets. For three or four successive nights those shapes might suddenly resolve themselves into destroyers with guns flaring, boats damaged, and torpedo hits claimed. Then there might be another fortnight of watchful silence.

It was not until the night of 8/9 November 1942 that *PT 61, PT 39* and *PT 37* slightly damaged the destroyer *Mochitsuki.* Five nights later *PT 40* and *PT 59* apparently interrupted a bombardment of Henderson Field, while a series of attacks by eight PT-boats on 7/8 December 1942 scored no hits, but persuaded a force of destroyers to turn back.

Two nights later the big 2001·5-tonne (1970-ton) submarine *I 3* was trying to supply Japanese troops in the Solomons when she was located in Kamimbo Bay, torpedoed and sunk by *PT 59.* Then on 11/12 December 1942, eleven Japanese destroyers were returning after floating supply canisters ashore at Cape Esperance when they were intercepted by three PT-boats. *PT 44* was sunk, but *Terutsuki* was set on fire by two torpedoes from *PT 37* and *PT 40.* Flames reached her depth-charge magazine and she blew up.

The PT-boats' attempts to repeat these successes were frustrated in January and early February 1943. The Japanese destroyers were determined and expecting this form of assault, while the growing number of PT-boats in the area needed more information from coastwatchers and air reconnaissance, and more coordination to be effective. Too many boats made their sighting signals while going into the attack, instead of waiting for their message to be acknowledged. Too often they lost touch before the word had been passed.

Catalina aircraft were now helping by illuminating the enemy with flares and by diverting his attention from the sea. There were a number of interceptions during this period, and many torpedoes were fired, but only *Hatsukaze* was damaged (by *PT 112* on 10/11 January 1943). *PT 112* was lost that same night together with *PT 43,* while *PT 111, PT 37* and *PT 123* were sunk on 1/2 February 1943. However, *Makigumo* could only evade the torpedoes fired at her by running into a minefield and sinking off Savo Island.

PT 66, PT 67, PT 68, PT 121, PT 128, PT 143, PT 149 and *PT 150* from Tufi in New Guinea played an important part in the Battle of the Bismarck Sea from 2 to 3 March 1943. They sank *Oigawa Maru*, one of the transports taking reinforcements from Rabaul to Lae, and were then engaged in the macabre aftermath of this battle.

Of the 8000 troops and naval personnel in these vessels, some had gone down in the ships while another 2734 had been rescued. The rest, many carrying small arms, were now in inflatable rafts and collapsible boats, or were clinging to bits of wreckage. New Guinea was very close and once ashore, they would have been formidable additions to the army facing the Allies in the jungle. Aircraft and PT-boats had no alternative but to circle the area, machine gunning and depth-charging the men in the water until none were left alive.

The PT-boats in the South and South-west Pacific, also engaged a number of inter-island barges, and were involved in the delivery and evacuation of Marine raiding parties. Their high speed and low silhouette made them ideal for this work, and MTBs were similarly employed in other areas, including the Aleutians. Fast fighting boats were not so suitable for the penetration of defended harbours, contrary to prewar expectations. Once surprise had been lost and their speed reduced, they proved flimsy targets, yet the mere fact of their availability necessitated their employment in such operations. Indeed the only Victoria Cross awarded to the Royal Navy's Coastal Forces was won during the raid on St Nazaire. The object of the expedition was the destruction of the entrance to the only Atlantic drydock big enough to take the German battleship *Tirpitz.* The lock-gates were to be rammed by the destroyer *Campbeltown* on the night of 27/28 March 1942. The Naval Force Commander's headquarters was in *MGB 314*, because she was fitted with radar.

Also present was *MTB 74,* one of the new Vosper 22m (72ft 6in) boats fitted with Packard engines, and specially modified to attack *Scharnhorst* and *Gneisenau* at Brest. A jumping wire running the length of the boat enabled her to slide under light surface obstructions, heavier ones being cut through or bounced over. Her two torpedo tubes had been moved to the foredeck, each being equipped with extra powerful compressed air launching cylinders. The two torpedoes, without engines but each fitted with a warhead of 500kg (1100lb) of ammonal would be thrown through the air for some 50m (50yd), clearing any further booms and splashing down to explode underneath the German capital ships. But *Scharnhorst* and *Gneisenau* had now left Brest, and so *MTB 74* was attached to the St Nazaire raid. She was allocated several targets, the final choice depending upon the situation on the night.

Campbeltown, MGB 314, and *MTB 74* would be accompanied by sixteen MLs. Most of them carried commandos, although four were fitted with conventional torpedo tubes for use against shipping in the harbour.

One ML had to abandon the mission en route, and twelve others were set on fire and lost during the action. *ML 160* was able to fire a torpedo at an enemy vessel and later escaped. *MTB 74* (commanded by Sub-Lieutenant R. C. M. V. Wynn RNVR) launched her special torpedoes at the lock-gates of the submarine basin. Withdrawing at full speed, she stopped to pick up more survivors and was fatally hit. *Campbeltown* rammed her objective and *MGB 314* (under Lieutenant D. M. C. Curtis RNVR, but with Commander R. E. D. Ryder RN on board) was soon the last British vessel left afloat in the harbour. The gunlayer at her forward 2-pounder was Able Seaman W. A. Savage, and he had already silenced a flakship during the final stages of the approach. He now engaged a pillbox at 200m (200yd) range, and remained on the exposed foredeck coolly selecting further targets amid the crossfire that covered the harbour. He was killed by splinters from one of the last shells, having earned a (posthumous) Victoria Cross, four more such awards going to other forces in this action. *MGB 314* was scuttled after reaching the supporting destroyers in the open sea. Next day, *Campbeltown* blew up, wrecking the dock. The day after that *MGB 74*'s torpedoes exploded.

Extra fuel tanks on deck had enabled the MLs to reach St Nazaire and – the survivors – to return under their own power. *MGB 314* and *MTB 74* had had to be towed for most of the way; and this was how the Norwegian *MTB 56* had crossed the North Sea back in October 1941. She was one of the 22·9m (75ft) Thornycrofts, powered by four of that company's 484·9kW (650bhp engines), two being coupled to each shaft. She had a speed of 29 knots and carried two torpedoes. Her captain was Lieutenant P. E. Danielsen RNorN, who, the previous month, had accompanied Lieutenant-Commander Pumphrey in a combined, though uncoordinated, MTB-MGB attack on a German convoy. Lieutenant Danielsen therefore had this experience behind him on the night of 3/4 October 1941, when he found the Norwegian tanker *Borgny* in Korsfjord. Evading *M 1101* and *V 5505, MTB 56* torpedoed and sank *Borgny*, then met the old torpedo boat *Draug* to be towed back to Scapa Flow.

Other attempts to provide ocean-going range, included the transport of small MTBs on board Allied and German warships. *LS 2 (Leichte Schnellboot*

2), *LS 3* and *LS 4* were three German MTBs carried respectively by the three raiders *Komet, Kormoran,* and *Michel.* Stowed in one of the holds, these 12·5m (40ft 10·3in) boats had a range of 200 miles at 22 knots and could carry four mines. They were of some use for inshore reconnaissance and minelaying, but did not fulfil their expectations, the raiders' success depending upon disguise rather than a high speed boat.

HMS *Fidelity* was a Free French Q-ship, who would pretend to straggle astern of a convoy, tempting a U-boat to surface and finish her off with gunfire. *Fidelity's* hidden armament would then be revealed, while *MTB 105,* lowered into the water on the disengaged side, would circle round and depth-charge the U-boat if she dived. *U 435* did not surface, but torpedoed *Fidelity* in the Atlantic. *MTB 105*, a tiny Thornycroft of 9·1 tonnes (9 tons) and a length of 14m (45ft 9in), floated off and drifted for two days. Sighted by the corvette *Woodstock* on New Year's Day 1943, *MTB 105* was despatched as a danger to navigation.

It was perhaps ironic that *Komet,* one of these MTB-carrying raiders should herself be destroyed by a high speed attack craft. For her second break-out, *Komet* was escorted down-Channel by torpedo boats and mine-sweepers. Mines had already taken a toll of four R-boats when the Germans were engaged by groups of MTBs and Hunt-class destroyers. *MTB 236* was one of the latest Packard-engined Vosper 22m (72ft 6in) boats, but she had lost touch with her force. Sub-Lieutenant R. Q. Drayson RNVR soon saw signs of battle ahead and then sighted *Komet* making good her escape. *MTB 236* crept in to 500m (500yd) from her bow before firing and racing away to starboard behind a smoke-screen. Both torpedoes hit and *Komet* blew up so violently that the shock damaged *MTB 236*'s engines. There were no survivors from the German ship.

Such dramatic actions undoubtedly had an effect on the general course of the war, as well as encouraging one side and disheartening the other. Yet most results were being achieved against ordinary coastal convoys, with S-boat and MTB flotilla commanders now considering how best to improve their performance. From April 1942 onwards, all S-boat flotillas came under the *Führer der Schnellboote, Kapitän-zur-See* (Captain) Rudolph Petersen. He decided on the priorities to be accorded to boats in the Black Sea, the Mediterranean, the Baltic, the North Sea, and the English Channel. Having his headquarters at Scheveningen in Holland, he was particularly well situated to arrange flexibility of operations in the last two areas.

The Germans generally favoured a minelaying campaign along coastal convoy routes, the S-boats sometimes screening R-boats and sometimes laying mines themselves. When it was judged that the British convoys had been lulled into a sense of false security or their escorts had been weakened, then one or two S-boat flotillas staged a series of concentrated raids on that route, ending abruptly as German attention was turned elsewhere. On 24 November 1941 five boats attacked a convoy off the East Coast, and again five nights later. On 11 and 15 March 1942 it was the English Channel, and in July 1942 Lyme Bay. On 2 October 1942 four S-boats struck at a convoy near the Eddystone, but five nights after, attention was focused on an assault by nine boats off the East Coast. Again they came thither on 14 October and on 9 November 1942. Another flotilla was in action off Plymouth ten days after that, and there were two more Channel sorties (and one on the East Coast) in December 1942. Some thirty-six merchantmen were sunk or damaged in all these attacks.

The Royal Navy, with its traditional commands of The Nore, Dover, Portsmouth, and Plymouth, preferred to send out some boats every night so that the enemy's coastal routes had no respite. As this meant that there was less chance of catching him unawares, officers like Lieutenant P. G. C. Dickens RN, Lieutenant C. W. S. Dreyer RN, and Lieutenant E. D. W. Leaf RNVR, had to consider various ways of achieving tactical surprise. The stealthy approach was all-important, and at last Dumbflow silencers were being fitted to all boats enabling them to make a quick approach on main engines and then accelerate away just as quietly. If the torpedoes missed, the MTB could get away undetected without alerting the enemy and spoiling another boat's attack. Even if the torpedoes hit, the enemy might suspect a mine if he heard and saw no attacker. Once detected, the boat could create a diversion, while an accompanying group of MTBs approached from a different direction.

MGBs could also keep an alerted enemy occupied, while the MTBs stole up on his disengaged side. MGB commanders often found themselves in an ideal position for launching torpedoes which alone could sink ships. So, there were growing demands for torpedo tubes to be mounted in MGBs, while MTBs received more powerful gun armament to smother or distract the escorts. Eventually the two types merged, all being known as MTBs, but coordinated tactics could only yield results after many disciplined exercises, informal discussions, and official reports. Each command now had a Coastal Forces Staff disseminating information, and ensuring technical support and maintenance, and an adequate supply of spares. The old and unreliable Mark V torpedo

Vosper 21.3m (70ft) boats at Felixstowe, *MTB 29* on the left. The officer standing on the quay is Lieutenant P. G. C. Dickens. *Imperial War Museum*

which could only carry its 227kg (500lb) warhead at a speed of 35 knots, was replaced by the Mark VIII*. This torpedo had a speed of 45 knots and a 340kg (750lb) warhead and could be armed after only travelling 91·4m (100yd).

Coastal Command struck at convoys during the day, reporting their position so that MTBs could attack after dark. But even narrow waters seem very wide on a dark night with visibility down to 200m (200yd). Some idea of the difficulties and monotony experienced by Coastal Forces on routine, uneventful patrols, can be gained from the fact that Lieutenant-Commander Hichens carried out 148 sorties, only fourteen of which resulted in action. A few degrees out in laying a course in a boat hammering so hard that one coxswain dislocated his spine; a momentary snuggle into quilted oilskins, because nothing had been seen for the past month and nothing would be seen tonight; a few seconds to wipe binoculars or readjust one's eyes after glancing at a dim light; and the target could be missed. It also worked the other way: a merchantman's lookout who stared at an exploding tanker, or relaxed as his ship turned into the harbour channel . . .

At the same time both sides were improving their defences against high speed attack craft. Both increased their mercantile armament. The Germans established standing trawler patrols, heavily armed with 8·8cm, 7·5cm, 40mm, and 20mm guns. Four of these vessels in a diamond formation could bring most of their guns to bear on any quarter and they dealt roughly with any MTB within range. In their turn, they became targets for torpedoes, as did the trawlers and corvettes employed on similar duties along the English coast.

S-boat victims also included the destroyers *Vortigern, Penylan* and *Eskdale,* while *Lightning* was sunk in the Mediterranean and British MTBs accounted for the torpedo boats *Iltis* and *Seeadler.* Equipped with radar, heavier calibre guns for longer range, and numbers of close quarters weapons, plus the advantages of speed and stability in rough seas, both sides' destroyers usually proved tough opponents, and drove off their small assailants on most occasions. This happened on 24/25 January 1943 when *Mendip* and *Windsor* tackled sixteen S-boats near Lowestoft and turned them back. *Montrose* and *Garth* hit one group of minelaying S-boats so effectively on 17/18 February 1943 that *S 71* stopped and was sunk by *Garth*. On the night of 7/8 March 1943 *S 114* and *S 119* collided while escaping the destroyer *Mackay.* It had become customary for MGBs to patrol the S-boats' routes across the North Sea and Channel, often right up to the hostile bases. Thus *S 119* was found and sunk by *MGB 20.*

Special air patrols were set up, Spitfires and Typhoons sinking *S 75* off Ijmuiden on 5 March 1943. British MTBs could be just as vulnerable to determined air attack, as was demonstrated on 14 September 1942 when the Italian 13th Fighter-Bomber Group sank *MTB 308, MTB 310* and *MTB 312* returning from the Tobruk raid. They also damaged *MTB 314* so badly that she had to be beached and was later taken into German service as *RA 10.*

To defend themselves, the S-boats received 40mm, 37mm, and 30mm dual-purpose weapons, in addition to their original 20mm guns. The British counterpart to this armament was mounted in the Fairmile D-class, and eventually comprised two 6-pounders, two 20mm, and eight machine guns, with two 53·3cm (21in) torpedo tubes in most. Slower than the planing 'short boats', the range of the Fairmile D-class and its seagoing ability, and accommodation were much enhanced, the 'Dogs' being big enough to carry the latest radar and communications equipment. The high speed attack craft of all navies were ready for the last years of the war. Their aim was now clear: the prevention of the enemy's coastal traffic in areas where he had no suitable alternative.

11 THE LOW KILLERS

The Baltic, Norway, the North Sea, the English Channel, Sicily, the Ligurian Sea, the Adriatic, the Aegean, the Black Sea, the Pacific; in all these areas the Allies took the offensive, amphibious assault being their principal mode of advance. Sometimes they came in massive invasion fleets; sometimes in a single fishing boat. Before each landing, information had to be gathered, mines swept, and the enemy's warships removed. Wherever waters were shallow or constricted, this work fell almost entirely upon coastal forces, the fast fighting boats either undertaking it themselves or protecting slower, more specialised craft. Often they cooperated with friendly aircraft, but just as often they were alone under hazardous skies.

Their hostile counterparts had the task of attacking the invasion forces and of warding off the Allied MTBs, who were trying to disrupt the Axis coastal traffic. In each area, the enemy had to reinforce and maintain his garrisons ready for each inevitable onslaught. Aircraft bombed his factories and land communications. Commandos and partisans ambushed his transport columns. Every delay when cargo was transhipped from barge to lorry to railway waggon to railway waggon of different gauge to lorry to mule, increased the risk of aerial destruction, sabotage and pilfering. How much more secure if the consignment could be stowed aboard one ship and not touched until it reached its destination. In some parts of the world there was an abundance of islands, or the local roads were so primitive that they could not even bear peacetime traffic; there, any movement had to be by sea. The Axis had few big merchant-ships left now and it would have been uneconomic and foolish to use them to carry a mere hundred crates to some remote jetty. Local barges, sailing vessels, fishing boats, and ferries would be perfectly suitable, and so would surplus minesweepers and redundant landing craft. They could easily hide from air attack during the day and continue their journey at night. They would be indistinguishable from the partisans' and commandos' own raiding craft.

If intercepted, their shallow-draught would save them from torpedoes and their 40mm, 20mm, and machine guns would be as heavy as anything mounted in MTBs. If they were hit, it would be no great loss, for they would be so close inshore that they could be run aground and most of their cargo saved. Indeed the boat herself might be patched up later and returned to service.

Such warfare seemed to have little of the spectacular success expected in prewar days. A caique set afire here, or a barge run aground there, lacked the drama of an exploding battleship, and frustration was compounded with every sinking, for then there were fewer targets for the growing number of Allied

MTBs. Those that did still float, concealed themselves longer and more effectively. Although in the end, through prolonged delay, some besieged strongpoint ran out of ammunition and surrendered, the MTB crews were not to know that. But by now, all theatre commanders appreciated the value of their coastal forces and gave them due credit for their contribution to victory.

Only the Japanese had neglected this aspect of naval warfare. They tried to make good their deficiency, but it was getting too late. Their basic design was derived from the MAS-boat acquired in 1940 and rebuilt as an MGB in 1941. Known as the H-1 (Hayabusa) type, her dimensions were 18m/18·7m overall × 4·5m × 2·1m (59ft/61ft overall × 14ft 9in × 7ft). She carried two 20mm, two 7·7mm guns and two depth charges. Of similar dimensions and comparable performance with two 45·7cm (18in) torpedoes, were the six MTBs of the T-1 type, built at Tsurumi in 1941. Sister-ships were ordered in that year, but the T-23, T-25, T-31, T-32, T-33, T-34, T-35, T-37, T-38 and T-39 types did not begin to arrive until late 1942 and 1943. Diversions and protection during attacks on enemy warships would be provided by a parallel development of MGBs: the H-2, H-35, and H-38 (of steel) types. The H-61 MGB type was also of steel, but had a waterline length of 19·0m (62ft 4in) and like many of her predecessors, mounted three 25mm guns.

The 31·5m (104ft 4in) T-51 type of MTBs were built in the later stages of the war and displaced 76·2-81·2 tonnes (75-80 tons). *No 10* carried four 45·7cm (18in) torpedoes, but the other seven only had two. The smallest Japanese MTBs were the T-14 and T-15 types with a length of 15-15·5m (48-50ft), and a displacement of 15·2 tonnes (15 tons). The number of Japanese boats was augmented by the Thornycroft CMB *Q III* captured in the Philippines, and by several Dutch boats of the TM-4 class. Five of these had tried unsuccessfully to intercept a Japanese destroyer force in the Dutch East Indies on the night of 19/20 February 1942. Next month they were all scuttled at Sourabaya, while those under construction were blown up. The Japanese later salvaged some of them using such allocated materials as were still available to build others of the type.

Altogether 353 high speed attack craft were built for the Imperial Japanese Navy. The chief problem faced by their crews is indicated by this list of the powerplants installed in them:

two 373·0kW (500bhp) Thornycroft; two 671·4kW (900bhp) Isotta Fraschini; two 671·4kW (900bhp) Type 94; three 671·4kW (900bhp) Thornycroft; three 335·7kW (450bhp) Lorraine Dietrich; three 354·4kW (475bhp) Kermath Sea Raider; two 783·3kW (1050bhp) Kansei; one 335·7kW (450bhp) Type 91; one 686·3kW (920bhp) Type 71; two 298·4kW (400bhp) Hispano-Suiza; two 373·0kW (500bhp) Hispano-Suiza; two 335·7kW (450bhp) Type 91; two 522·2kW (700bhp) Lorraine; two 686·3kW (920bhp) Type 71; two 320·8kW (430bhp) Type 3; two 447·6kW (600bhp) Type 2; two 522·2kW (700bhp) Type 41; two 522·2kW (700bhp) Kinsei; two 537·1kW (720bhp) Type 21; two 223·8kW (300bhp) Type 51; two 1342·8kW (1800bhp) Vulcan; four 671·4kW (900bhp) Type 6; one 686·3kW (920bhp) Type 91.

It is hardly surprising that they were plagued with engine troubles and that their top speed could be anything from 17 to 40 knots. Captain Tameichi Hara, senior instructor at the torpedo school, said of one, 'This is no torpedo boat. It is nothing but a barge.'[1] He was immediately forced to preach the doctrine that any success could only come through stealthy approach, not through dramatic speed and combat. But even those vessels which reached operational areas were restricted through lack of petrol. Most were employed on harbour duties and none seem to have been able to interfere with the American PT-boats' campaign in the South and South-west Pacific.

So far the PT-boats had been equipped with the old Mark VIII torpedoes originally intended for destroyers in the 1920s. They did have a range of 914·4m (10,000yd), but as the boats lacked fire control over this distance and were usually much closer, they were saddled with torpedoes that were heavy, slow (27 knots), and with a mere 136kg (300lb) warhead. The Mark VIII torpedoes had to be kept upright and had to be launched from tubes to avoid tumbling the control gyros. The black powder launching charge often revealed the PT-boat's position at night, and sometimes set fire to any lubricating oil still in the tube. A compressed-air launcher was needed, its development being undertaken by Higgins Industries. Now the 45-knot Mark XIII aircraft torpedo was arriving, more reliable and with a 272·4kg (600lb) warhead. It also had a non-tumbling gyro, so that it could be just slid from a rack, thus saving weight. Some PT-boats dispensed with torpedoes altogether, while all increased their gun armament with 37mm, 40mm, and even 75mm weapons. Although comparatively light, the 75mm was not so popular as those of smaller calibre because of its slower rate of fire and its lack of tracer, so that the gunner could not follow the line of his shells. Some boats were equipped with a 20mm four-gun powered mounting, the Elco Thunderbolt. Other boats had two 12-round 11·4cm (4·5in) rocket projectors, followed by the 12·7cm (5in) fin-stabilised rocket. Depth-charges could also be dropped close alongside the target to break her back.

With few big warships in PT waters, this armament was employed against the barges being used to sneak supplies into Japanese garrisons on bypassed islands. The barges were armoured and carried 40mm guns, but they were slow and their only real protection was to remain hidden under the shore. Every night PT-boats patrolled among the rocks and islets, one of which might turn out to be a barge or coaster. The action usually ended with the target running aground and the PT-boat being driven off by shore artillery and mortars. In one year in one area, fifty barges and 150 smaller vessels were claimed destroyed, although some could well have been put back into service later. More PT-boats were lost through stranding and being scuttled to avoid capture in these poorly charted waters than through any other cause.

There were some famous episodes as on 16 September 1944, when Lieutenant A. Murray Preston USNR took *PT 489* and *PT 363* into Wasile Bay at Halmahera. Zigzagging across a minefield, they survived three hours of artillery fire until, 200m (200yd) from shore, they rescued a United States Navy pilot in a rubber dinghy. Lieutenant Preston was awarded the Congressional Medal of Honor. The dramatic Battle of the Surigao Strait (24/25 October 1944) again showed the difficulties experienced by MTBs attacking a well-composed battle-fleet. All the PT-boats involved early on were driven away and even the disabled cruiser *Mogami* could still deal roughly with any boat that tried to finish her off. Only two torpedoes hit that night, one from *PT 137* crippled the cruiser *Abukuma* and one from *PT 323* struck the destroyer *Asagumo* who was lying stopped. It was a different story on 11/12 December 1944, when *PT 492* and *PT 490* (both Elco boats) used their radar to locate a large target hove-to off Palomplon. Hidden by the shore, they crept up to 1000m (1000yd) and fired six torpedoes. Two hit the destroyer *Uzuki* and sank her. Again, on 26 December 1944, PT-boats harassed a Japanese amphibious assault on Mindoro after its repulse by army aircraft. While *PT 221* withdrew under heavy fire, *PT 223* made an unseen approach and torpedoed the destroyer *Kiyoshimo*. Already damaged from the air, she sank soon afterwards. Then on 2 March 1945, *PT 373* took General MacArthur back to Corregidor.

But for the most part, PT-boats acted as despatch vessels, hunted barges and prahus and burned isolated warehouses and dumps. It was dangerous work in hazardous waters, and it seemed to have little effect on the eventual outcome of the war and merely prevented remote garrisons becoming troublesome. Yet in this small way they were fulfilling one of their prewar expected roles, by releasing destroyers for service in the forefront of the advance.

Islands also featured prominently in the Mediterranean Campaign, and so did the establishment of mobile bases which were continually moving forward so that MTB patrols could keep ahead of the army's flank. Even then boats had to return to Malta for major repairs, such as deck strengthening to bear the weight of new guns. Longitudinal steel stiffeners lessened the risk of breaking the boat's back over a year of continuous service. Shafts became twisted and one flotilla needed 102 new propellers in two months, through striking submerged wreckage off invasion beaches.

First there was Sicily, which brought an unexpected encounter between three Vosper 22m (72ft 6in) MTBs and two surfaced U-boats proceeding to their patrol zone through the Straits of Messina. *MTB 81* (with the Senior Officer, Lieutenant Christopher Dreyer RN, controlling his flotilla as an admiral directs his fleet), *MTB 77* and *MTB 84* were lying stopped and listening when *U 375* was sighted 300m (300yd) away. *MTB 81* just avoided her by going astern and had only backed some 100m (100yd) when *U 561* also passed ahead. A torpedo hit the U-boat and she blew up with two survivors. *MTB 77*'s torpedoes misfired, and *U 375* dived, thus avoiding the torpedoes from *MTB 84*. The night was not over for the MTBs were engaged by shore batteries and fired a torpedo at a group of E-boats, which were later intercepted by *MTB 655*, *MTB 656*, and *MTB 633*. Two E-boats were destroyed by the Fairmile 'Dogs'.

MTB 422 **is a 24.7m (81ft) Higgins boat (ex** *PT 92)* **made available to the Royal Navy in 1943; she is seen passing wrecked shipping at Livorno.** *Imperial War Museum*

On some Mediterranean nights, special PT-boats broadcast the noises of a full-scale landing, diverting the enemy's attention from the real assault on some other beach. Coastal Forces were also involved in the abortive attempt to hold Kos and Leros, and later were the only Allied warships operating in the Aegean. Much of the traffic was carried in caiques which were often boarded and sunk with demolition charges.

Meanwhile, other MTBs, MGBs, and MLs had moved into the Adriatic. In spite of a damaging air raid on the main base at Bari, an advanced base was established on the island of Vis under Lieutenant-Commander Morgan Giles RN. Coastal Forces cooperated with Yugoslav partisans, raiding islands and disrupting inshore traffic so effectively that the Germans were forced to occupy several islands preparatory to an assault on Vis itself. The campaign continued, the largest warship in the zone (the old cruiser *Niobe*) being wrecked by *MTB 298* and *MTB 226*. In twelve days, Lieutenant-Commander T. Fuller RCNVR captured eight schooners, one lighter and one motor boat, and sank one schooner, one motor boat, one tug and one tanker – all by keeping a few yards offshore, before surprising the enemy. The Admiralty said that 'The tactics evolved date from the 18th century and earlier, and are evidently as effective now as then'.[2] The German assault in Vis was called off as its retention could not be guaranteed if such a campaign of interdiction continued; as indeed it did, up into the northern Adriatic. But targets were becoming fewer. One group of MTBs carried out sixty-four patrols with only one action. On the other hand one sortie by four Fairmile MTBs destroyed eleven vessels and damaged another four. Their Senior Officer (Lieutenant-Commander T. J. Bligh RNVR) reported that it 'was made possible due to low visibility, land background, uncertainty of identification, absurdly close ranges, excellent gunnery, and admirable coolness'.[3]

Somewhat different tactics were being employed on the west coast of Italy where German convoys were protected by F-lighters *(Fährprahm)* and flakships armed with 8·8cm anti-tank guns. The sea was too open for the MTBs to effect surprise, so the 8·8cm kept them out of torpedo range and easily outdistanced the lighter Coastal Forces' weapons. Commander Robert Allan RNVR accordingly formed a miniature battle squadron of three landing craft (gun) (LCG), each armed with two 4·7in destroyer guns manned by Royal Marines. They would deal with the F-lighters, enabling six Fairmile MTBs and MGBs to get in close and attack the convoy. The LCGs could only do 10 knots, so they were screened by two Higgins PT-boats, while three others scouted ahead.

The whole fleet was controlled by Commander Allan in *PT 218,* whose PPI (plan position indicator) radar screen gave a clear picture of what was going on. He was accompanied by another Higgins PT-boat as reserve 'flagship'. Evaluation of the radar plot and the skilful direction of a well-trained and cohesive, yet individualistic, force achieved several notable successes. These were enhanced by the arrival of new non-contact pistols, which detonated torpedoes as they passed under the target and were particularly effective against the shallow-draught F-lighters. In fact, on the night of 12/13 April 1945, Lieutenant-Commander C. Jerram RNVR and the Vosper 22m (72ft 6in) *MTB 410, MTB 409* and *MTB 408* sank five F-lighters with six torpedoes fired on radar without visual sighting.

A similar campaign was being conducted in the winding Norwegian fjords and leads. This time the base was Lerwick in the Shetlands and the Fairmile Ds were the only boats to have the range, seaworthiness and accommodation for extended North Sea operations. Generally, the long boats had difficulty in achieving surprise in the open sea, but their size was not such a disadvantage in a rocky inlet. They could also lie up overnight awaiting targets of opportunity. For camouflage, they were furnished with nets big enough to cover the whole boat and of a reversible design, to simulate rocks and trees on one side and snow on the other. A telephone cable linked the boat with an observer on a mountain top overlooking the nearest channel. The necessary planning and mounting of these operations was the responsibility of Lieutenant P. D. Job RNVR because of his personal knowledge of the Norwegian coast. Moon, weather, and operational circumstances were not always favourable, so he was allowed to make all the arrangements at the Lerwick base. When the boats were underway Lieutenant Job informed Admiral Sir Lionel Wells (Admiral Commanding Orkney and Shetland), who confirmed their orders and sent the necessary instructions to other units. For example, Bristol Beaufighters usually provided escort halfway across the North Sea, and met the MTBs on their way back. Information about shipping movements came from aircraft, fishing boats, and agents.

The 30th (Norwegian), 54th (Norwegian) and 58th MTB Flotillas all served in these waters. On the moonlit night of 27/28 November 1942, two MTBs stole up on two merchant ships anchored at Askevold. Torpedoes were launched at 700m (700yd); one ran wild, one hit the 5080·0-7112-tonner (5000-7000-tonner), and two exploded in the 6096-8128-tonne (6000-8000-ton) Taurus-class vessel. Both sank, while one of the trawler escorts fired three tracer shells as the MTBs withdrew at 25 knots. For five days in January 1943 *MTB 618, MTB 619, MTB 620, MTB 623, MTB 625, MTB 626, MTB 627,* and *MTB 630* struck at German installations in Sognefjord and Korsfjord. Commandos were put ashore on Stord to wreck a copper mine, a quay was destroyed with torpedoes, prisoners taken, and a Ju 88 shot down. *MTB 619* and *MTB 631* experienced one of the hazards of this area when the latter ran aground during a sortie off Florö on 13/14 March 1943. She had to be abandoned, but the raid did account for a 1269·0-tonne (1249-ton) merchant ship out of a three-ship convoy. The MTBs continued minelaying and torpedo operations even during the short summer nights. On 4 June 1943, the 8262·1-tonne (8132-ton) *Altenfels* was sunk with four torpedoes and *M 468* raked with gunfire near Traelso. The latter replied, killing two and wounding five men aboard *MTB 620* and *MTB 626.*

The Dog-boats were the most heavily armed light craft of the war. This Fairmile D is bristling with two automatic 6-pounders in power turrets fore and aft, twin 0.5in machine guns either side of the bridge, twin 0.303in machine guns on the bridge, a twin Oerlikon amidships and two 53.3cm (21in) torpedoes. *Imperial War Museum*

In the last two winters of the war, the MTBs formed part of a combined air, surface, submarine and sabotage offensive against shipping off the Norwegian coast. *MTB 627* and *MTB 653* sank *Irma* and *Henry* near Kristiansund on 13 February 1944. *MTB 712, MTB 722,* and *MTB 711* wrecked the coaster *Freikoll* near Florö on 8 October 1944. Sometimes the defences proved too awake, as when *V 5101* and *V 5113* protected their Stavfjord convoy. At the beginning of November 1944 *MTB 712* and *MTB 709* sank *V 5525* and *V 5531* in Sognefjord, but on 13 November 1944 *UJ 1430, UJ 1432,* and *V 1512* drove off *MTB 688* and *MTB 627* in the same area. Although *MTB 715* and *MTB 623* failed to achieve any success against *V 5514, V 5527* and *RA 203,* that same night (27/28 November 1944) saw the destruction of the 5542·3-tonne (5455-ton) *Welheim* by *MTB 717* and *MTB 627* who withdrew after being hit by the escorts.

The merchant ship *Ditmar Koel* was sunk by *MTB 717* and *MTB 653* on 8 December 1944, while on 23 December 1944 *MTB 722* and *MTB 712* sank the minesweeper *M 489* in Bömslofjord. Three days later *MTB 717* and *MTB 627* destroyed the tanker *Buvi* in a small convoy off Fröysjoen.

One operation involved MTBs taking four Welman craft into Bergen. These one-man submarines (built at Welwyn Garden City in Hertfordshire) approached their target in surface trim with the pilot's cupola just above the water. They then dived below the enemy's hull to release a charge with magnetic attachment. During the operation, one Welman was captured and the other three had to be scuttled. Another idea was that of the boom patrol boat or airborne attack boat, seventeen being built by Vosper. It was 5m (16ft) in length. A Lagonda V12 engine drove an outboard unit with coaxial contra-rotating propellers, further directional stability being provided by two plate fins. Dropped by parachute in a fjord, the boat would be steered around to a neighbouring anchorage where the pilot would aim for his target and escape on a life raft. The boat would run on and be detonated by a bumper rail in the bow. These craft never saw service, but conventional MTBs undertook 152 Norwegian operations, sinking twenty-five enemy ships at a cost of eighteen men

A German patrol ship moves along the Norwegian coast unaware of two Fairmile Ds which lie in wait (at the lower right side) in the small harbour at the bottom of the cliff – an indication of the value of small craft for miscellaneous duties in coastal water. *Liddell Hart Centre/Lt Cdr Dalzel-Job*

killed and fifty wounded. Many of their battles were fought against the weather and the Fairmiles proved they could survive, although few experienced the ordeal of *MTB 625* and *MTB 666.* Separated in a Force 11-12 storm, they could only make 1-2 knots over the ground. Both took on so much water that they had to jettison their torpedoes. Both reached shelter, but *MTB 625* had a mere 15cm (6in) of freeboard forward, while the water in *MTB 666* was reaching the spark plugs on her last serviceable engine.

These Norwegian MTB operations were temporarily halted in the summer of 1944 in preparation for the Normandy landings. As 1943 wore on into 1944, both sides knew that the invasion must be coming and the fast attack craft's warfare intensified as both sides strove for nocturnal domination of the English Channel and southern North Sea.

The S-boats were receiving new torpedoes during these months. The T5 or *Zaunkönig* used a hydrophone to home on to the noise made by a ship's propellers, being particularly effective against escort vessels. The FAT *(Flächenabsuch Torpedo)* started zigzagging after a certain distance, a fatal manoeuvre to ships in convoy. The LUT *(Lage unabhängiger Torpedo)* circled after its run, eventually becoming a floating mine of great hazard to shipping in a crowded anchorage. The slow 9 knots of the *Dackel* (Dachshund) gave it a very long range of 35½ miles. Some *Dackel* were carried by S-boats although they were really intended for coastal defence, being launched from shore-mounted tubes. The S-boats continued to use the old G7A, a compressed-air torpedo with a warhead of 380kg (836lb). Its speed fell off from 44 knots over 3¾ miles to 30 knots over 8½ miles. The later G7E which saw service for most of the war had a 500kg (1102lb) warhead and a range of 3¾ miles at 30 knots, without leaving a track of compressed-air bubbles to reveal its approach. Both the G7A and the G7E could be detonated by contact or magnetic pistols.

The S-boats employed all these weapons, plus mines, against coastal convoy routes for the rest of the war, switching between areas in the hope of catching escorts off guard. Sometimes they were successful, as on the night of 27/28 April 1944 when nine attacked an invasion rehearsal convoy of eight LSTs in Lyme Bay. Through various mischances the sole escort comprised the old destroyer *Saladin* and the corvette *Azalea. LST 507* and *LST 531* were sunk, while *LST 289* was damaged, 638 soldiers and sailors being lost.

More often the S-boats were located and driven off by destroyers, then intercepted by Coastal Forces. Indeed, after D-Day (when improved maintenance ensured that 97 per cent of allocated MTBs were operational) Coastal Forces' controllers were established ashore or were carried aboard certain frigates and destroyers. Watching radar screens, they coordinated their MTB groups in blockading S-boat ports and in repelling those that did manage to break out. When necessary, the bigger warships themselves joined in, for no longer did fast attack craft exchange light gunfire harmlessly. Now there was hardly an action without at least one boat being badly damaged or sunk. Belts of 2-pounder ammunition included starshell, originally for illumination but capable of starting fierce fires in wooden vessels.

Airborne cannon and rockets, too, were more precise than the old bombs and machine-guns. Waves of fighter-bombers pressed home their attacks on S-boats at sea. Ninety-seven B-17s and B-24s bombed their construction yards at Bremen-Vegesack on 18 March 1943. On 26 March 1944, 358 B-26s struck at Ijmuiden but the concrete pens were strong and only *S 93* and *S 129* were destroyed. Some of the heaviest bombs were dropped on the night of 14/15 June 1944 when 325 Lancasters attacked Le Havre. For some reason the R-boats were using the concrete bunkers with the result that *S 84, S 100, S 138,*

Lürssen's *S 170,* built in 1944.

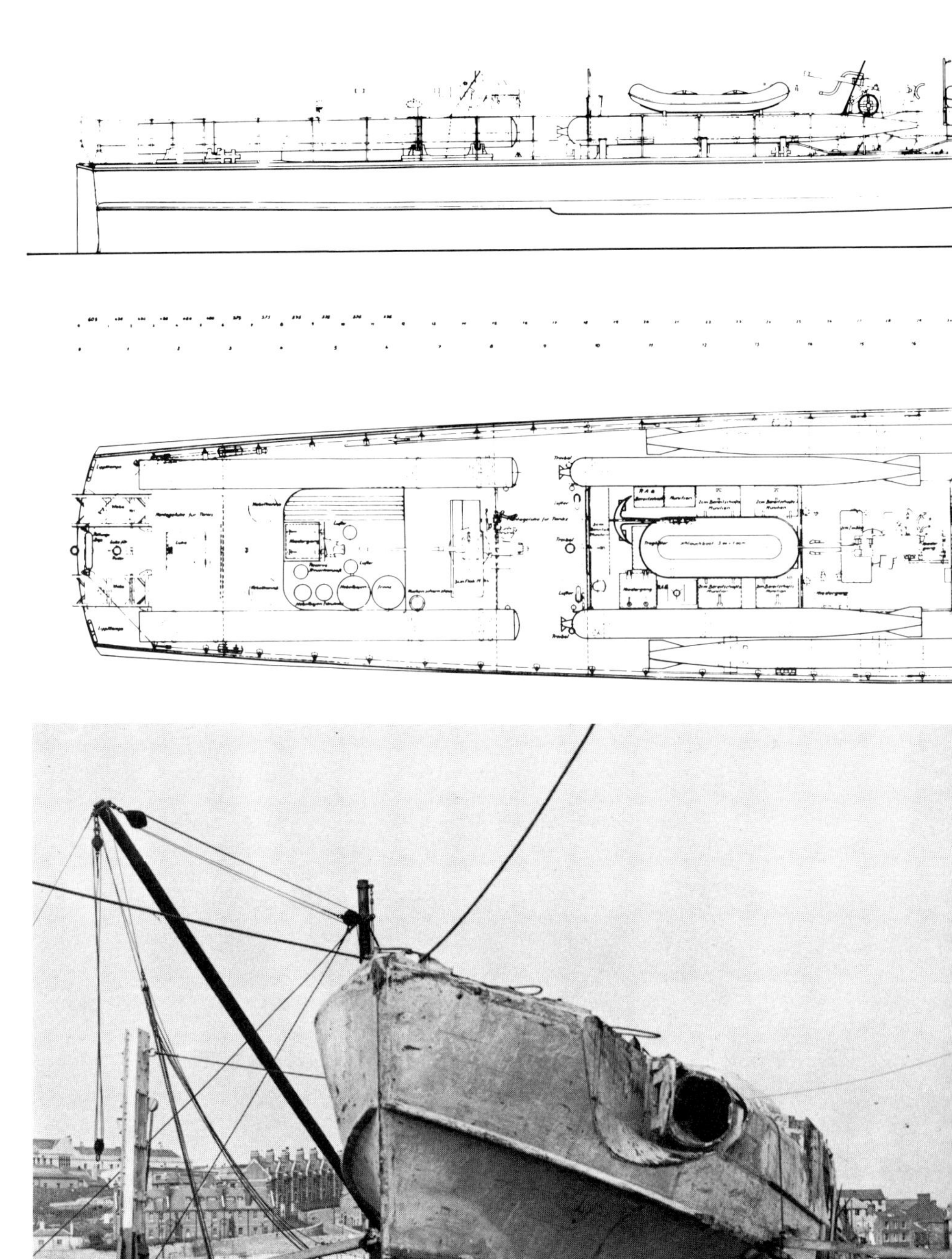

A captured S boat at Plymouth in 1945. *Imperial War Museum*

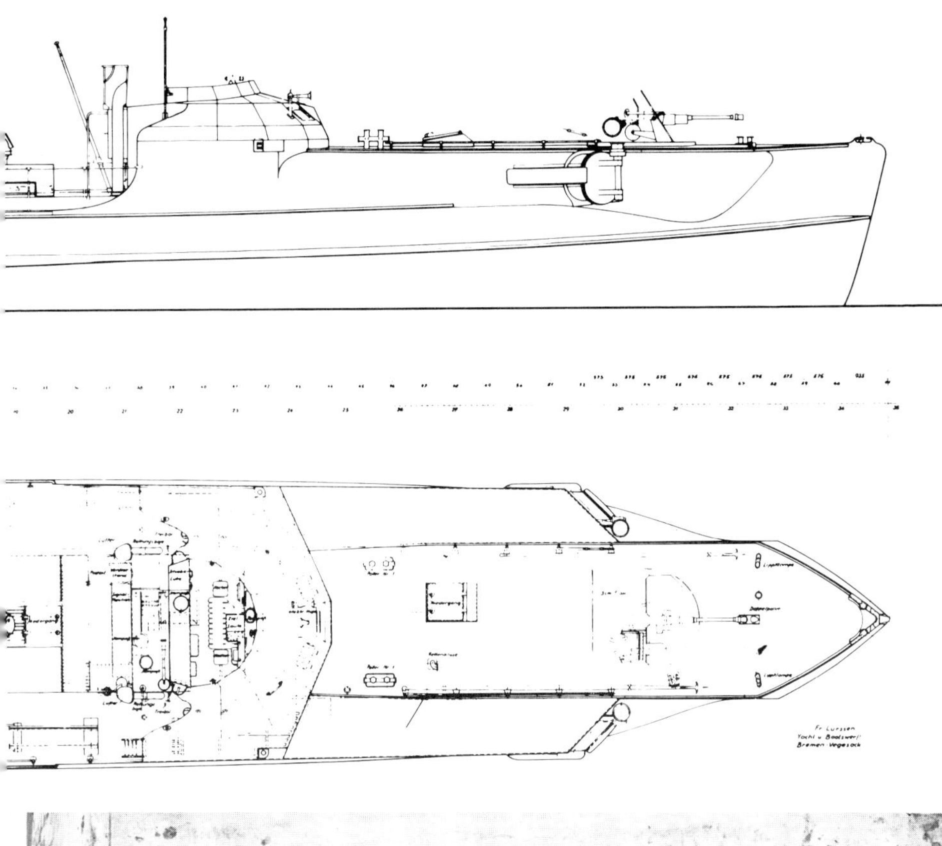

Stern gear of the same boat, showing the main and auxiliary rudders and the trim wedges fastened below the transom.
Imperial War Museum

S 142, S 143, S 144, S 146, S 150, S 169, S 172, S 173, S 177, S 187 and *S 188* were all destroyed together with twenty-one other warships, plus a variety of harbour craft. This disaster was followed by an accidental explosion at the torpedo repair depot, forty-one torpedoes being destroyed, with a corresponding limitation of operations. Yet no matter what befell them, no matter what the weapons and tactics used against them, or how hard they were crowded and outnumbered, the S-boats were never entirely eliminated from the English Channel and North Sea.

Explosive motor boats (called Weasels by the Allies) had made their debut in the Channel on the night of 2/3 August 1944. The *Linse* (Lentil) was a 1·5-tonne (1·5-ton) motor boat with a range of 60 miles at 31 knots. The pilots of two boats steered their craft towards their target and then abandoned them. A third *Linse* took over radio guidance until the target was hit. The two pilots were recovered and the surviving boat withdrew at full speed. Twenty of these craft belonging to the 211th Small Battle Unit Flotilla under Lieutenant-Commander Bastian attacked shipping off the Normandy beachhead. Completely unarmed, they had no defence against the MTBs, MGBs and MLs hunting them down, but their first sortie was not the total failure that Allied information claimed it to be. The destroyer *Quorn,* the trawler *Gairsay,* and an LCG were all sunk, while the merchant ships *Fort Lac La Ronge* and *Samlong* were damaged.

However, no matter how effective the first operation was in such cases, subsequent attacks lacked the 'secret weapon' element of surprise and all were broken up. Winter attacks off the Belgian coast (including one employing S-boats as *Linsen* carriers) had to be abandoned because of adverse weather conditions. Light Coastal Forces also dealt with German human torpedoes, and both these and the *Linsen* saw further service in the Mediterranean. Usually they failed, although the Italian Fascist Naval Forces did achieve some surprise on 12 February 1945. The cruiser *Delhi* was in Split harbour in the Adriatic when she was the target of six explosive motor boats. Four reached the harbour but three were destroyed by *Delhi's* guns and shore batteries. The survivor hit a landing craft alongside the cruiser and blew up some 20m (20yd) away, shaking the 4927·6-tonne (4850-ton) warship considerably.

Although the pilots of all these small craft sometimes rode them in until they hit, they were not designed as suicide boats. The *Shinyo* ('Ocean Shakers') were. These 5-5·5m (16ft 6in-18ft) boats were powered by one or two motor-car engines, giving a speed of 26-28 knots. The majority were made of plywood, although a few were of steel. They were armed with two 12·7cm (5in) rockets which would be fired by the pilot to spray exposed gunners with incendiary bullets, but their main load was a mass of high explosive, or two depth-charges. About 6000 *Shinyo* were built, 2000 of which were soon in position around the coasts of Japan ready for launching against the expected invasion fleet. The others that had left their factories were distributed among various parts of the shrinking empire. They were first encountered in Lingayen Gulf on 9/10 January 1945, when a group of *Shinyo* sank *LCI(M) 974* and *LCI(G) 365. LST 1028, LST 925,* and the transport *War Hawk* were damaged, and so was *LST 610* the following day. *PC 1129* was sunk near Nasugbu also in the Philippines on 31 January 1945.

The attack cargo ship *Starr* was damaged by a *Shinyo* off Okinawa on the night of 8/9 April 1945. Meanwhile the destroyer *Charles J. Badger* was firing at shore targets near Kezu Saki. Just after the change of watch at 4.00am, while it was still dark, an engine was heard and then an explosion burst into the starboard side of the after boiler-room and flooded through into the after engine-room. The starboard shaft was jammed; the radar and much other equipment wrecked. *Charles J. Badger* had to be towed to Kerama Retto for repairs, but the *Shinyo* escaped after dropping its depth-charge.

The task of warding off these attacks was usually assigned to PT-boats, who also investigated rocky coves and overgrown inlets, searching for the little green boats. They destroyed more than 100 *Shinyo* in the Philippines and succeeding campaigns, but *PT 77* and *PT 79* were themselves mistaken for these deadly craft. Fired on by American destroyers in a reef-strewn area near Talin Point, Luzon, one ran aground and the other had to reduce speed. Both were hit by shells and blew up.

PT-boats also came under attack from aerial Kamikazes, although their high speed dodging usually saved them. The PT-tender *Oyster Bay* and her brood shot down three suiciders on 26 November 1944. On 5 December 1944 *PT 494* was dived on by a Zero which followed her movements precisely. *PT 494* and her sister Elco *PT 531* continued firing until the last moment, when *PT 494* turned to port and the Zero hit the sea 25m (25yd) to starboard. A similar manoeuvre saved *PT 532* five days later, but *PT 323* was not so fortunate and was practically cut in half. In anchorages, the PT-boats usually kept between the Kamikazes and the valuable transports, firing as long as the enemy was in the air – and picking up survivors when the enemy did get through. Sometimes the onslaughts were mixed with conventional bombing and strafing, and sometimes the PT-boats were themselves singled out deliberately. Within one and a half minutes three aircraft dived on the Higgins *PT 230* who zigzagged to keep her starboard side towards the nearest plane. With her guns blazing, she turned hard-a-starboard right under the diving aircraft which struck 9m (30ft) away. The second attacker missed by 15m (50ft), but the third crashed right by *PT 230*'s stern, lifting her out of the water, shaken but unscathed. *PT 77* stopped dead just before the moment of impact, so that the fourth suicider missed ahead by 9m (30ft). *PT 223* turned to starboard and avoided a fifth Kamikaze by 3·0m (10ft), while two more straddled *PT 298* in quick succession just 5m (15ft) and 1m (3ft) away. The final aircraft was shot down by combined fire from the

Japanese *Shinyo* explosive motor boats in Picnic Bay after the recapture of Hong Kong, 1945. Rails and trolleys aid rapid launching from their lair. *Imperial War Museum*

boats. That was on 16 December 1944. Next day *PT 74* and *PT 84* survived similar attacks, but the day after that a Kamikaze followed *PT 300*'s final turn to starboard and blew her in half. Further damage and casualties were inflicted during the seemingly unending days of Kamikazes, but no more PT-boats suffered direct hits.

By now Russian fast attack craft had had wide experience in cooperation with their advancing armies. In spite of deficiencies in radar and communications equipment, their torpedo cutters had seen action in the Black Sea, the Baltic, and the Arctic. In all these theatres they had harassed German convoys, fought off S-boats, torpedoed larger warships, and landed commandos. They had not always met with the successes enjoyed by similar craft in other navies, but they had certainly worked very closely with the Red Army. Now, in the Far East, forty-five former PT-boats were among the 204 torpedo-cutters taking part in the Soviet invasions of Korea, Sakhalin, and the Kurile Islands in the last days of the war. Coastal traffic was attacked and troops landed.

Far to the south, on 30 August 1945, British ships were again entering Hong Kong harbour. As they did so, three *Shinyo* entered the water from their lair at Lamma Bay. They, and their shore installations, were destroyed by aircraft from the carriers *Indomitable* and *Venerable.* High speed attack craft had fought their last action of the Second World War.

12 ENGINES, HULLS AND WEAPONS

Quite suddenly high speed attack craft were neither useful nor glamorous. They were small and uncomfortable. They could not handle minesweeping gear and did not make economical harbour craft or liberty boats. Most of them needed refits, having been driven hard away from permanent base facilities. Hulls under repair in forward areas had often transferred generators, radios, and pumps to operational vessels so that most units had at least one cannibalised boat. But even if they had been renovated, their limited range, seakeeping ability, and accommodation did not suit them to the flag-showing duties of peacetime; there were enough surplus destroyers and frigates to do that. Nor was it likely that MTBs would influence anybody's foreign policy in a world dominated by the atom bomb. Besides, most admiralties look askance at 'private navies' and that was how many sections of the world's press had portrayed light coastal forces. So, it is not surprising that the Allied navies soon demobilised their MTB crews or transferred them to other duties. The boats were sold, some to foreign navies, the majority, disarmed, to civilian buyers.

Some were too far gone even for that. In the Pacific, 118 PT-boats were suffering so badly from damage, worm and rot, that they were de-equipped and burnt at Samar. The training MTB Squadron 4 was the last to be decommissioned on 15 April 1946. Only the 24·4m (80ft) Elcos *PT 613, PT 616, PT 619* and *PT 620* remained in the United States Navy's Operational Development Unit. There had been some American research during the Second World War. In December 1943 the standard 24·4m (80ft) Elcos *PT 487, PT 560, PT 561, PT 562* and *PT 563* were fitted with a six-step kit. The Elcoplane gave speeds of 53·6-56·0 knots and the boats could reverse course in 6 seconds, but they also increased their cruising fuel and lubrication consumption by 25 per cent and 75 per cent respectively. At cruising speeds, too, they were difficult to handle. The Elcoplanes and their associated hull fittings suffered damage in heavy seas and the idea was abandoned.

The Higgins *Hellcat*, on the other hand, was a completely fresh design. Numbered *PT 564* in 1943, this 21·3m (70ft) boat could average 47·8 knots and could reverse course in 9 seconds. Besides fuel economy, her small size had many advantages in a stealthy attack, but she was not large enough to carry a heavy armament and accommodate her crew in combat areas. No more *Hellcats* were built, although several companies undertook design work for PT-boat projects after the war.

In 1945 most of the Royal Navy's surviving MTBs were formed into four reserve flotillas, another two flotillas being retained for the training of young officers and for experimental work. Of course, the defeated nations surrendered their boats which were scuttled or treated the same way as the victors' vessels.

The Second World War had resulted in three main types of high speed attack craft: the hard-chine short boats (Vosper, British Power Boat, Elco and Higgins); the long boats, exemplified by the Fairmile Ds and by the S-boats; and the steam gun boats. Now, at last there was time for systematic examination of each type's advantages and deficiences.

Of them all, the SGBs had proved least effective in wartime. Admittedly, they were silent and well-armed, being considered formidable opponents by the S-boat crews, who often reported SGBs as 'destroyers'. But their large size prevented a surprise approach in the open sea and, with their steel construction, hindered the destroyer and frigate programmes. The SGBs' steam machinery (fuel tank to fuel pipe to boiler to steam pipe to turbine to steam pipe to condenser to water pipe to boiler – involving two separate compartments) proved much more vulnerable to close-range weapons than did the internal combustion boats with their basic fuel tank-fuel pipe-engine layout. The SGBs' machinery spaces had to be protected with armour plate, thus increasing displacement and reducing speed.

The first propellers installed in SGBs had to be redesigned as they suffered from cavitation, a phenomenon known since the early 1890s and frequently encountered by the designers of fast boats. As a propeller is turned, the water is screwed back past the line of the blade, thus thrusting the boat forward. The faster the propeller turns, the faster the boat goes. Eventually, more powerful engines turn the propeller so fast that the water does not have time to follow the line of the blade exactly, but begins falling behind it creating a virtual water vacuum. Not only is this inefficient, because the propeller is skidding and not producing so much thrust, but the aqueous vapour bubble or blister collapses on to other parts of the blade with a force equal in some cases to 1406kgscm (20,000psi). The movement of the propeller becomes rougher, while the blade itself is pitted, eroded, and eventually holed. One answer is a thinner blade, but that may become too weak. Another suggestion is to fit a wider propeller and slow it down with reduction gears, but then the blades interfere with each other, there is extra weight, and the engines' energy is wasted – the exact opposite of what was originally intended. The present practical solution is a very small fully-cavitating propeller mounted as deep as possible and turning so fast that the water bubble collapses well astern of the blade.

These supercavitating 'dryback' propellers are somewhat less efficient, because all the thrust only comes from the forward face of the blade, but they require thinner shafts and brackets and thus save weight. It took years of research in cavitation tunnels and on seagoing boats to establish these facts and their associated problems, such as repositioning rudders to prevent their erosion. Altering the pitch of the blades allows the propeller to function efficiently at both high and low speeds, but the necessary controls mean yet more weight. The search for the perfect propeller has not yet ended.

When considering the hull form of future MTBs it did seem that the best development would be along the lines of the German S-boats and the Fairmile Ds, both capable of fighting their weapons in Force 5 weather. Planing craft were limited to practical operations in Force 4 winds, although both long boats and short boats could survive in much worse seas. The Fairmile Ds had a range of 1200 miles at 10 knots and they were well-armed, but they were too slow, 30 knots being their best speed. The Fairmile F was a natural development, *MGB*

In 1946 Kris Cruisers launched the ASRL *Celerity*, built for the Bristol Aeroplane Company. She was similar to the Fairmile F and is seen here in the Solent in 1948.

2001 being the type's sole representative.

A similar vessel was completed after the war as an ASRL and named *Celerity*. Her four 1216·0kW (1630bhp) Bristol Hercules 14-cylinder HE-17 MB radial engines required very large air intakes for cooling and combustion, intakes which had to be fitted with special shuttering to keep out sea water. Designed by Fred Cooper and built by Kris Cruisers of Isleworth, Middlesex, *Celerity's* hard-chine hull was made of double-diagonal mahogany with an outer skin of hiduminium, a light alloy of aluminium, magnesium, and manganese. This material was also used for much of the superstructure. A speed of 40 knots and a cruising range of 4200 miles was expected during her trials following her launch in 1946. *Celerity's* dimensions were 32m × 6·7m × 1·8m (105ft × 22ft × 5ft 9in) on a displacement of 81·2 tonnes (80 tons).

Somewhat similar in appearance to the Fairmile Ds, but built in orthodox fashion and not prefabricated, were the Camper and Nicholson 35·7m (117ft) round-bilge boats with a range of 2000 miles at 11 knots. Several remained in service at the end after the war, three taking part in arcticisation trials in February 1949. Their guns were fitted with electric heaters, while their interior was insulated with glass wool 75mm (3in) thick. A number of the earlier Camper and Nicholson boats had been powered by three 746kW (1000bhp) Davey Paxman diesel engines, German success with high speed diesels interesting the Royal Navy in this type of machinery. Although the British engines did not give the required speed, Sir Roy Fedden's 1943 Committee did recommend lightweight diesels for the Royal Navy's postwar MTBs. The increased efficiency of cycle, which drove the engine longer for the same amount of fuel and with longer periods between maintenance, offset the heavier scantlings needed to bear the weightier diesels. Apart from these advantages, the risk of fire would be lessened, especially when refuelling. The Ostend disaster of 14 February 1945 was probably the worst of its kind. A film of petrol from cleaning tanks had spread across the surface of the water and became ignited, possibly by a cigarette. The fire spread rapidly with ammunition, warheads, and fuel tanks exploding. *MTB 255, MTB 438, MTB 444, MTB 459, MTB 460, MTB 461, MTB 462, MTB 465, MTB 466, MTB 776, MTB 789, MTB 791,* and *MTB 798* were all destroyed, the loss of life being equally heavy.

Two of the trio of Camper and Nicholson's durable 116.8-tonne (115-ton) MTBs at Gosport in February 1949, just before departure to 'the coldest spot they could find' in the Arctic, to test weapons and equipment in sub-zero conditions.

By 1946, it was recognised that steam was an unsatisfactory power source for high speed attack craft, petrol adequate, and diesel best. But now a new type of engine was becoming available – the gas turbine. One was first installed in *MGB 2009,* one of the Camper and Nicholson boats (ex-*MGB 509* and also known as *MGB 5559).* The two wing shafts retained their Packards, but a 1865kW (2500shp) Metropolitan-Vickers G2 gas turbine was fitted to a gearbox on the centre shaft. An automatic clutch allowed this propeller to idle through the water when the wing engines only were being used. The work and subsequent evaluation was carried out by HMS *Hornet,* the Coastal Forces base at Gosport. It was found that a particular advantage was *MGB 2009*'s ability to move at a minute's notice because the gas turbine did not require warming up.

Vosper then undertook the conversion of *Grey Goose* (ex-*SGB 9*), removing the steam machinery and substituting a specially designed marine gas turbine, the Rolls-Royce RM60. Rotol controllable pitch propellers provided manoeuvrability and astern movement.

Meanwhile Vosper were going ahead with their own programme of research and development. In 1943 they had produced the experimental *MGB 510* with a length of 30·6m (100ft 6in). By the end of the war their short boats were carrying 70 per cent more military load than in 1938, on basically the same hull form of 21·3m (70ft) length.

The Royal Navy's first gas turbine ship (seen here in 1947 as *MGB 2009,* later re-classified *MTB 5559)* was an 116.8-tonne (115-ton) Camper and Nicholson boat capable of 31 knots. Experience with this experimental installation was applied to the Bold class five years later.

Now Commander Peter Du Cane was authorised to design a vessel embodying wartime lessons and investigating short boat problems. The result was *MGB 538*. Her hull was 22·6m (74ft 6in) long overall, its compromise hard chine shape being derived from Vosper's ASRLs. The chine was much higher and the bows flared out, resulting in improved seakeeping and less spray thrown over the bridge at cruising and idling speeds. She was of all-glued laminated wood construction, instead of the planking being screwed to the frames, a method which had led to lines of splits and cracks. The exhaust was originally silenced by being led through a hollow rudder stock and blade, but this was not successful. The usual hand/hydraulic steering with Dumbflow silencers for the exhaust was eventually used in *MGB 538*. The increasing amount of electronic equipment requiring alternating current had meant the installation of DC/AC converters, a source of weight circumvented in *MGB 538* by incorporating AC and DC converters on the same shaft. Originally mounting one 4·5in and two 20mm guns, she later carried two Oerlikons and four 45·7cm (18in) torpedo tubes and was renumbered *MTB 1601*.

By now the Admiralty had been so impressed with the results of craft converted to gas turbine propulsion, that they specially designed two vessels for these engines. They had two shafts, each with one Metrovick G2 gas turbine of 3189·1kW (4275shp). A gearbox coupled each shaft into its own Mercedes-Benz diesel for cruising and manoeuvring. This system became known as CODOG (Combined Diesel Or Gas), the S-boat diesels being replaced by Napier Deltics when they were ready. The exhaust was taken up through two large side-by-side funnels, giving the class a distinctive appearance. They were 37·2m (122ft) overall with a displacement of 131·1 tonnes (129 tons). To ensure adequate evaluation, one was built with a hard-chine hull by J. S. White. Vosper's round-bilge hull proved the wetter, with spray getting into the air intakes. In spite of baffles and settling chambers, this could result in salt corrosion in the engines. The two craft were not very successful at speed but they did show the feasibility of this form of propulsion.

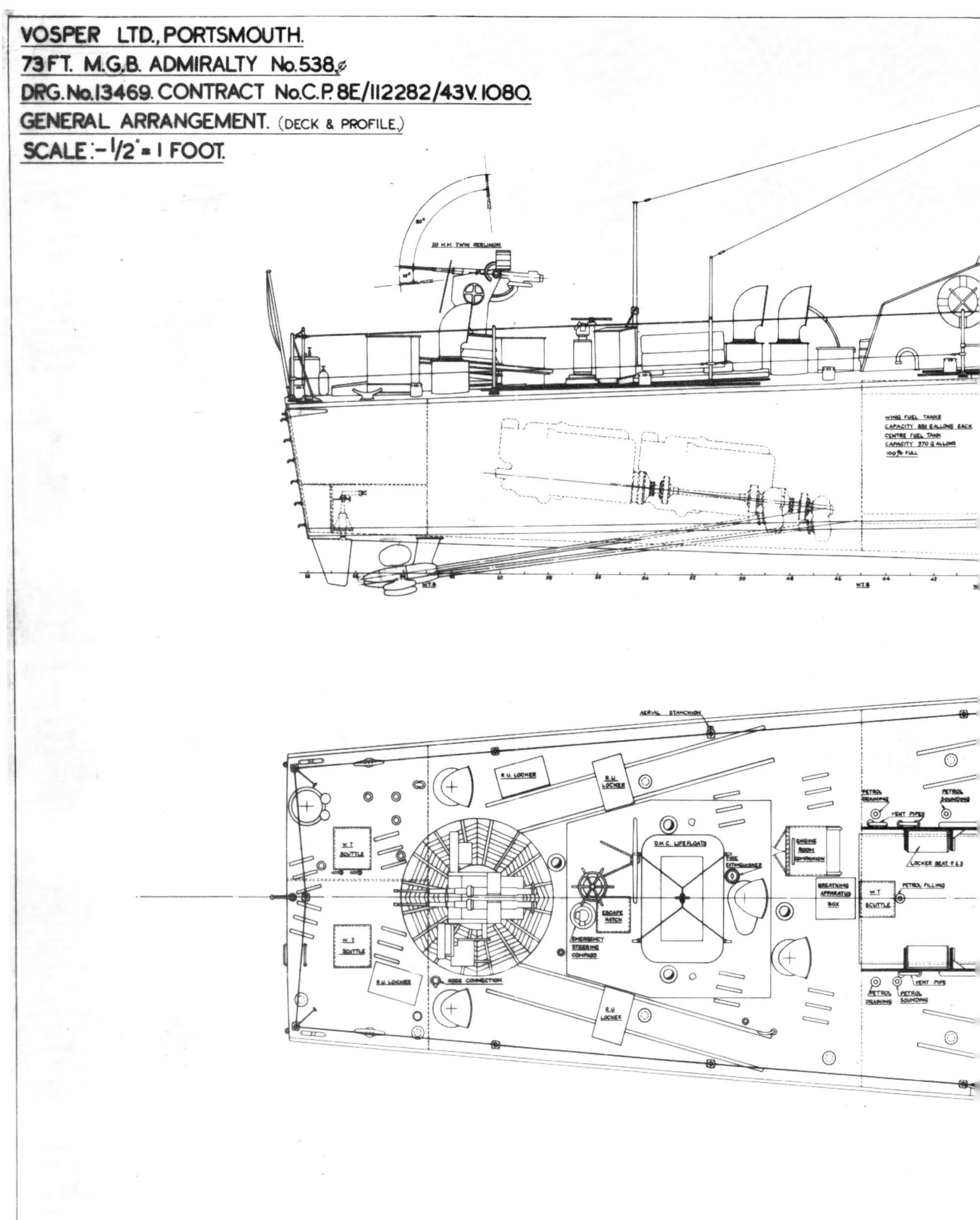

Vosper's experimental prototype of 1948, *MGB 538.*
Vosper

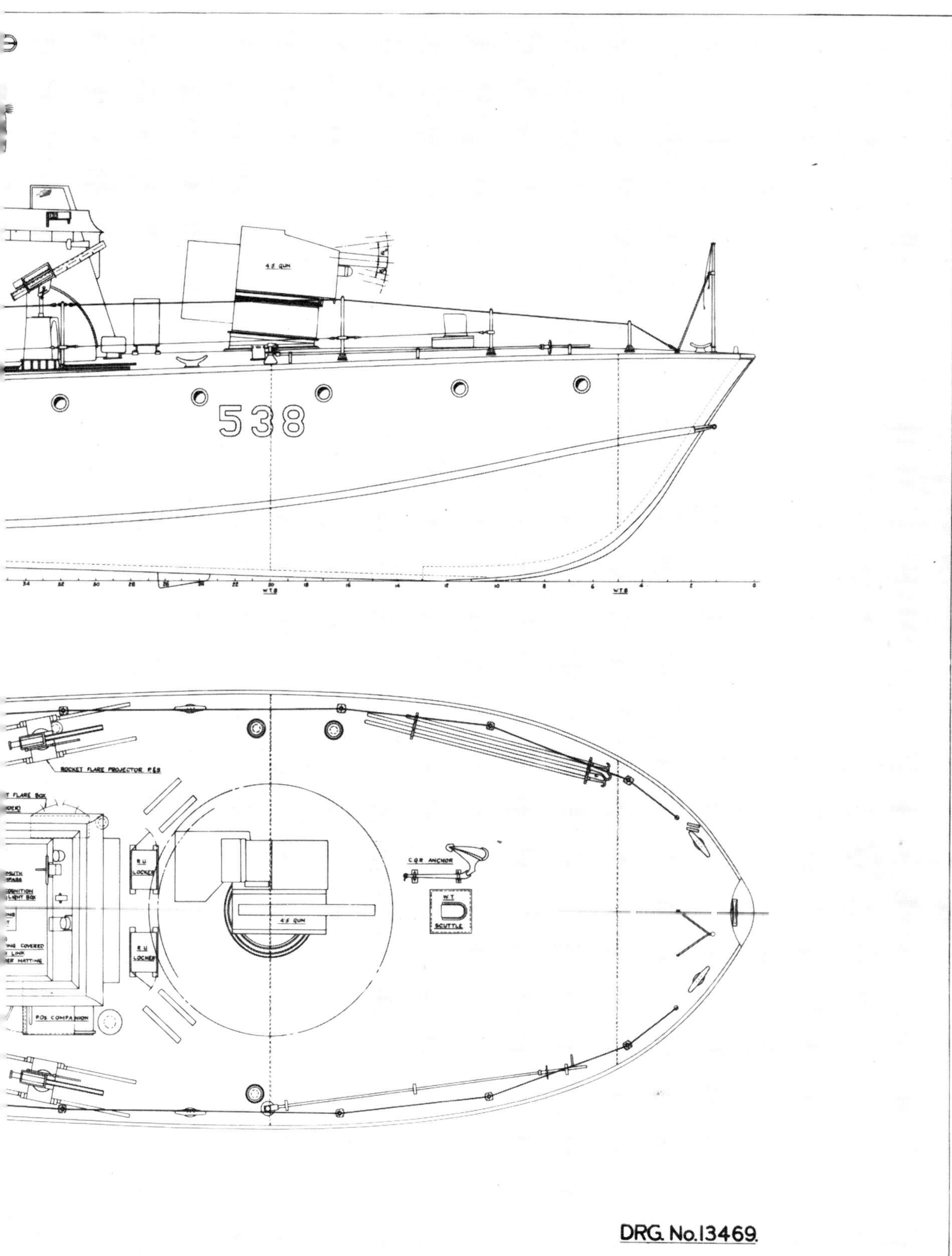
4.5 GUN
538
W.T.B.
W.T.B.
ROCKET FLARE PROJECTOR P.6.8
R.U. LOCKER
R.U. LOCKER
4.5 GUN
C.Q.R. ANCHOR
W.T. SCUTTLE
P.O.s COMPANION
DRG. No.13469.

VOSPER LTD., PORTSMOUTH.
73 FT. M.G.B. ADMIRALTY No. 538.
DRG. No. 13470 CONTRACT No. C.P. 8E/112282/43V. 1080.
GENERAL ARRANGEMENT. (ACCOMMODATION & SECTIONS)
SCALE:- 1/2" = 1 FOOT.

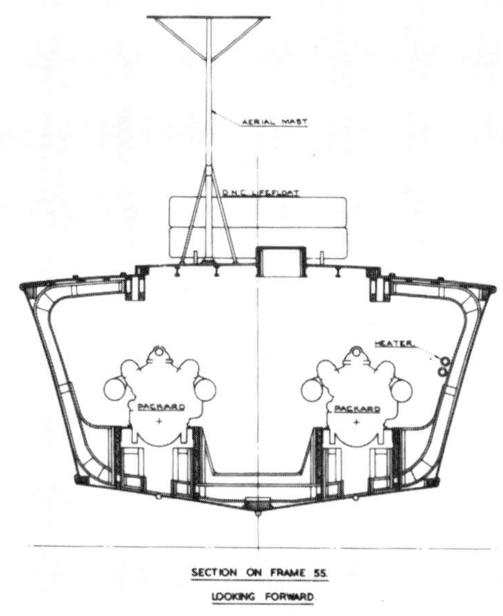

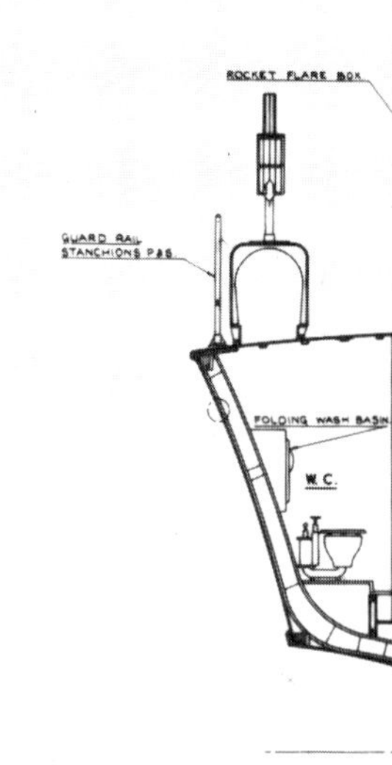

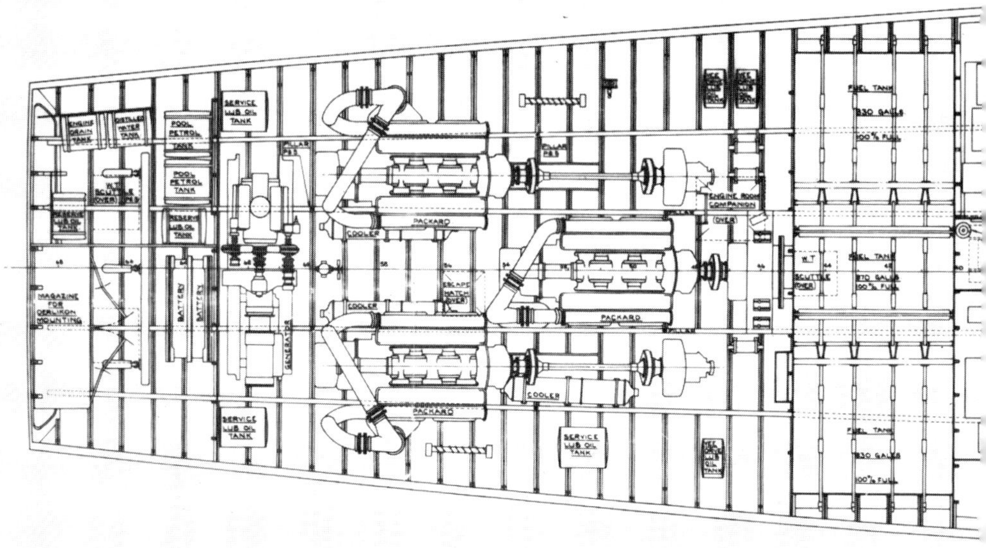

Vosper's experimental prototype of 1948, *MGB 538.* *Vosper*

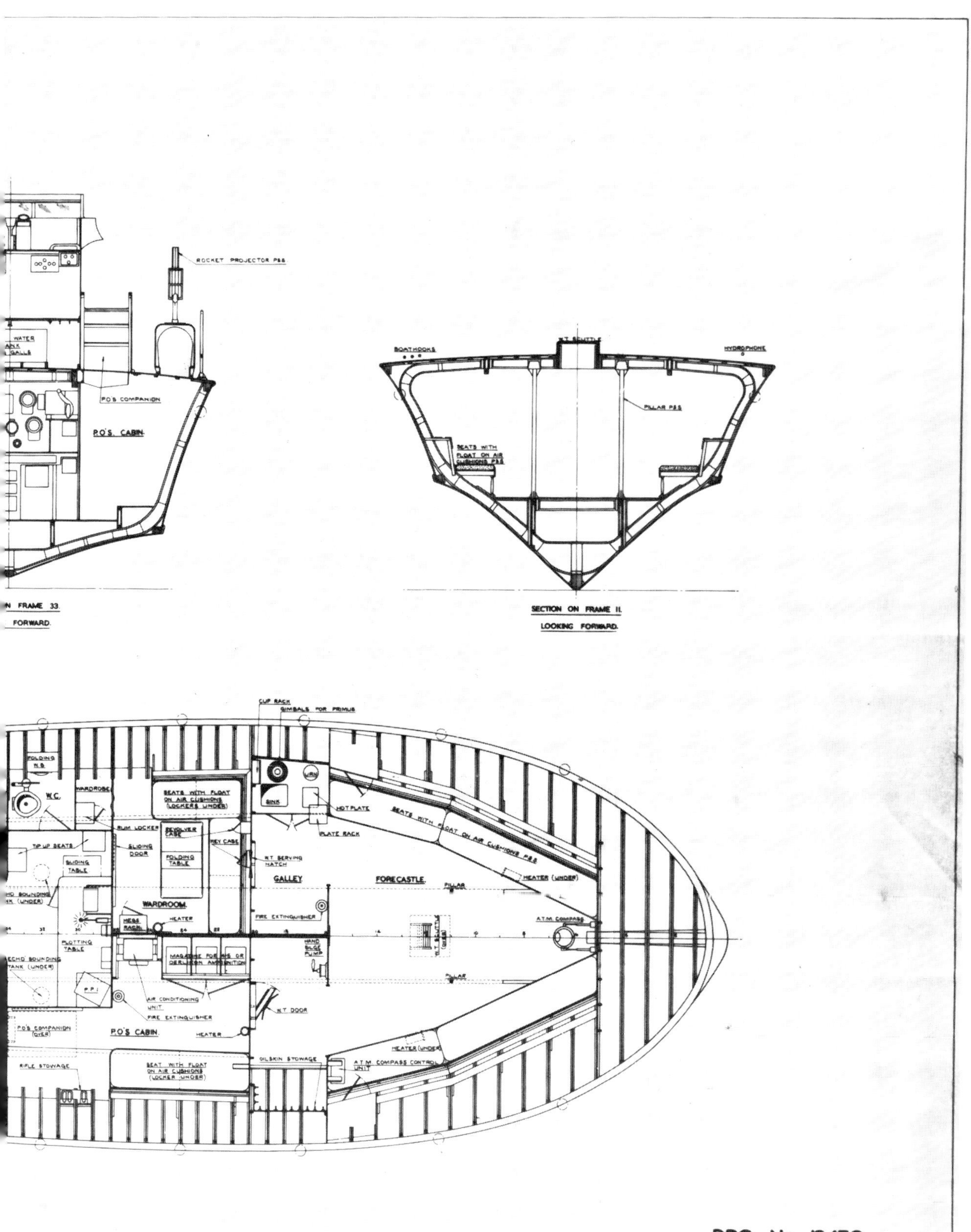
ROCKET PROJECTOR P&S
BOATHOOKS
HYDROPHONE
PILLAR P&S
P.O'S COMPANION
P.O'S. CABIN
SECTION ON FRAME 11.
LOOKING FORWARD.
CUP RACK
GIMBALS FOR PRIMUS
W.C.
SINK
HOT PLATE
PLATE RACK
RUM LOCKER
SLIDING DOOR
FOLDING TABLE
KEY CASE
GALLEY
FORECASTLE
PILLAR
HEATER (UNDER)
WARDROOM.
HEATER
FIRE EXTINGUISHER
A.T.M. COMPASS
HAND BILGE PUMP
PLOTTING TABLE
AIR CONDITIONING UNIT
W.T. DOOR
P.O'S CABIN
HEATER
OILSKIN STOWAGE
A.T.M. COMPASS CONTROL UNIT
RIFLE STOWAGE
SEAT WITH FLOAT ON AIR CUSHIONS (LOCKER UNDER)
DRG. No. 13470.

They also introduced two other novel features. The class was designed to be fully convertible. With a minimum of base assistance, their own crews could substitute minelaying equipment or a heavier gun for torpedo tubes. They could thus be employed in any role without having to carry the weighty armament of multi-purpose craft. To describe them as MTBs or MGBs could be misleading, for they might not always be carrying their designated armament. Nor, with gas turbines, could they be said to be powered by motors. Henceforth, all such vessels were classified as fast patrol boats, craft of less than 30 knots being known simply as patrol boats.

In fact, the principle of convertibility might not have proved so practicable on operations, as forward bases would then have had to hold complete outfits of armament in duplicate, besides stocks of the appropriate ammunition. In any case, wartime bases became adept at transferring equipment from damaged to serviceable boats. What was important was that FPBs should be designed to carry different weights of armament, even if they were not actually fitted.

Until now the Admiralty had regarded these vessels as mere boats, deserving only numbers and unworthy of names, crests, or trophies. Now that FPBs had apparently come to stay, their traditions could be handed on, and so could names. With a standard prefix to show their class *Bold Pioneer* was launched by White on 18 August 1951, *Bold Pathfinder* by Vosper on 17 September 1951.

These postwar developments and their associated exercises may not seem very significant, but they amounted to much more than many other countries' fast fighting boat activities. War-surplus MTBs were purchased by emergent nations but they lacked the prestigious appeal of an aircraft carrier or cruiser. Even the United States Navy ran down its fleet of PT-boats without replacement.

Thus, when the Korean War begun in 1950, none of the United Nations had sufficient high speed attack craft available for the campaign. On the other side there were four aluminium-hulled North Korean ex-Russian motor torpedo boats, armed with one machine gun each. On 2 July 1950 they were escorting ten coasters and trawlers down the east coast. Soon after dawn they were off Chumunjin when they encountered the cruisers *Juneau* and *Jamaica* and the frigate *Black Swan*. The four boats immediately attempted a torpedo attack, but the first salvo of proximity-fused (VT) shells blew up one boat and set a second on fire. Another ran aground and was destroyed by gunfire, while the survivor made off and seems to have played no further part in the hostilities. From then on until 27 July 1953, the naval war became one of bombardment and blockade.

The latter prevented the Communists from using coastal waters for transport and for fishing. All contraband was seized, the crews made prisoner, and the captured vessels destroyed or handed over to the South Koreans. Once officially declared, the blockade had to be rigorously enforced to be effective. To be legal, every part of the blockaded coast had to be inspected by ship every twenty-four hours.

This burden fell mainly upon the United Nations' destroyers and frigates, the smallest vessels being the 284·4-tonne (280-ton) patrol craft (PCs) of the Republic of Korea. Among the islands, estuaries, mudflats, rocks and bays, hostile sampans were rounded up, junks intercepted, raiding parties landed, a sack of rice or a cow captured for a friendly village. One night the 2788·9-tonne (2745-ton) destroyer *Nootka*'s radar located a spasmodically fast vessel with floating objects coming from her. She proved to be an oared junk, being used as a silent minelayer. The objects were the crew, armed with submachine-guns and floating in truck tyres. Danger from mines sometimes prohibited the

J. S. White's *Bold Pioneer,* built in 1951 to develop the potential of gas turbines. Like her sister ship, *Bold Pathfinder,* she was a dual rôle FPB, armed with either two 4.5in and a single 40mm gun, or with the 40mm mounted on the foredeck and four 53.3cm (21in) torpedo tubes. *P. A. Vicary*

employment of warships within the 10-fathom line and even occasionally within the 100-fathom line. Because of these and other restrictions, many vessels sent their ship's boats close inshore, arming them with small arms, bazookas and recoilless rifles. Lurking among the islets, they captured sampans, wrecked fishing nets, shot up jetties and directed naval bombardments. But they were mere improvisations, their greatest disadvantage being lack of speed.

It is not surprising that the Korean conflict should have hastened the development of high speed attack craft. Aluminium-hulled PT-boats powered by four Packard engines began trials with the United States Navy in 1951. Elco's *PT 809* was 29·9m (98ft) long with a beam of 7·9m (26ft). *PT 810* from Bath Iron Works (Maine) measured 27·1m × 7·3m (89ft × 24ft), while *PT 811*'s dimensions were 28·6m × 7·6m (94ft × 25ft). She was built by John Trumpy and Sons of Annapolis. Philadelphia Naval Shipyard produced a 32m × 6·4m (105ft × 21ft) hull for high speed in rough weather, but *PT 812*'s calm water speed was disappointing and she was later equipped with gas turbines. All four boats served as MTB Squadron 1 in the Operational Development Force from 1954 to 1959. Armament research included the consideration of rocket weapons.

The Royal Navy, too, produced Admiralty requirements and construction was put in hand. First in service were the Type B or Gay-class, virtually Admiralty repeats of Vosper's earlier short boats. Built of wood they had an overall length of 22·9m (75ft) and a beam of 6·1m (19ft 6in). They displaced 46·7 tonnes (46 tons), with three 1212·3kW (1625bhp) Packard petrol engines giving a speed of 42 knots. Transom flaps were fitted retrospectively, to adjust the trim and thus keep the bow up in a following sea (to improve stability) and the bow down in a head sea (to reduce slamming). *Gay Archer* was launched on 20 August 1952, being followed by *Gay Bombardier, Gay Bowman, Gay Bruiser, Gay Carabineer, Gay Cavalier, Gay Centurion, Gay Charger, Gay Charioteer, Gay Dragoon, Gay Fencer,* and *Gay Forester.*

Dark Hunter was the first Type A to be launched on 18 March 1954. She was built by James Miller of St Monance (Fife), although the design work had been done by the Admiralty and the parent firm of Saunders-Roe at Beaumaris (Anglesey). Like their prototype *MTB 1602* (formerly *MTB 539),* the Darks were constructed of aluminium alloy frames with wooden skin and deck.

After the war one of the S boats assigned to Britain was used as a test-bed for the Napier Deltic. This remarkable but fickle engine was an 18-cylinder diesel of triangular configuration which produced 1865kW (2500bhp): current versions of the Deltic, now built by Paxman, produce 2984kW (4000bhp) and are in widespread use. ***P 5212*** **is seen in 1953 during the naval review at Spithead.**

Their overall dimensions were 21·7m × 5·9m × 1·9m (71ft 4in × 19ft 10in × 6ft), with a full load displacement of 71·1 tonnes (70 tons). They mounted one 40mm gun, but the rest of their armament could comprise one 4·5in gun or four 53·3cm (21in) torpedo tubes or six mines. Their powerplant was quite different from the three Packards in *MTB 1602*. At last a suitable British diesel was available being derived from a specification proposed by the Fedden Committee. The Napier Deltic had eighteen cylinders arranged in an opposed triangle with three crankshafts. Light alloys kept the weight down to some 4·6 tonnes (4·5 tons), and the Deltic produced 1865·0kW (2500bhp). Development had begun in 1947, a completed engine running in 1950. After sea trials in 1952 two Napier Deltics were installed with V-drives in each Dark boat. It was hoped that they would do over 50 knots, but the boats proved overweight when completed. Restricted air intakes and exhausts, unsatisfactory propellers and vibration, prevented the employment of full engine power, limiting their best speed to 46 knots, with 37-42 knots more typical.

Dark Adventurer, Dark Aggressor, Dark Antagonist, Dark Avenger, Dark Biter, Dark Buccaneer, Dark Clipper, Dark Fighter, Dark Gladiator, Dark Hero, Dark Highwayman, Dark Hunter, Dark Hussar, Dark Intruder, Dark Invader, Dark Killer, Dark Rover, and *Dark Scout* were completed between 1954 and 1957. The last-named was built entirely of aluminium alloy as was a group of five sold to Burma at a total cost of £1,800,000. These particular boats were equipped with Saunders-Roe slow-speed electric drive for service in

A dozen Vosper designed Gay class boats were completed in 1953-4, all powered by Packard V12s. Some of the class (though not *Gay Bombardier,* above) were laid up on completion, without even being launched. After they were sold on the surplus market, one of the boats set sail for rebel Biafra, but was interned at Gibraltar.

intricate waterways among Burmese islands and rivers. *PT 9* of the Japanese Maritime Self-Defence Force was another Dark-type sold abroad. *Dark Attacker, Dark Battler, Dark Bowman, Dark Chaser, Dark Chieftain, Dark Crusader, Dark Defender, Dark Explorer,* and *Dark Horseman* were also on order for the Royal Navy but they were cancelled after the end of the Korean War. As a further economy the Royal Navy's Coastal Forces base at HMS *Hornet* was closed down. The older boats were sold and all the newer ones put into reserve except for three fast patrol boats (FPBs) forming a Trials and Special Service Squadron at *Dolphin II* (the old *Hornet* berth at Gosport).

Yet British knowledge of fast fighting boat construction and operation did not entirely lapse. In 1955 thc Admiralty placed a contract with Vosper for a design study for a new type of FPB. The principal specifications required a 50-knot vessel capable of carrying torpedoes and a 3·3in Coastal Forces 2 (CFS2) gun weighing 8·1 tonnes (8 tons). Vosper's proposals were accepted and two prototypes were ordered in 1956: *Brave Borderer,* launched on 7 January 1958, and *Brave Swordsman* launched on 22 May 1958. They were of composite construction; welded aluminium alloy frames, bulkheads and deck, sheathed with double-diagonal mahogany planking secured by stainless steel bolts with plastic insulating bushes to prevent electrolytic corrosion. The hard-chine planing hull form was that proved by *MTB 1601*'s seakindliness, its waterline and between perpendicular lengths being 29·3m and 27·4m (96ft and 90ft) respectively. Overall dimensions were 30·1m × 7·8m × 2·1m (99ft × 25ft

6in × 7ft), with a full load displacement of 115·8 tonnes (114 tons). The superstructure was of aluminium alloy and fibreglass. Development of the CFS2 gun had been cancelled, so the torpedo boat armament comprised one 40mm/L60 Bofors and four 53·3cm (21in) torpedoes. As torpedo gunboats they mounted two 40mm guns and two 53·3cm (21in) torpedoes. The torpedoes were carried in light racks, further alterations permitting employment as gunboats (two 40mm guns), minelayers (one 40mm gun and ten mines), or raiding boats (one 40mm gun and two inflatable outboard dinghies).

The heart of the design was made up of three Bristol Siddeley Marine Proteus gas turbines, each weighing 1·4 tonnes (1·3 tons) and developing a maximum continuous power rating of 2480·5kW (3325shp). The free power turbine (on a separate shaft from the compressor and its turbine) was chosen because its interaction with the propeller at varying depths is more flexible than that between a diesel and its propeller. A special gearbox was designed, combining V-drive, clutch, reverse gear and reduction gear, one being fitted to each shaft. Experiments in Vosper's own cavitation tunnel indicated that small supercavitating propellers would be best. The weight saved by shrinking these on to their shafts would offset their reduced efficiency. A range of 400 miles at maximum continuous power was provided by 25·4 tonnes (25 tons) of fuel. Separate Rover gas turbines drove the electric generators, insulation and earthing ensuring a minimum of interference with radio and radar equipment. The machinery was controlled by one man outside the engine room.

Saunders-Roe built the experimental prototype of the Dark-class *(MTB 1602)* in 1950. This boat was powered by three Packards but the rest of the class, 1954-8, were fitted with Napier Deltics.

In 1960, Du Cane piloted *Brave Borderer* through her most significant trials in a Force 6 wind. His experience of rough sea conditions and the use of transom flaps resulted in a speed of 52 knots, a convincing demonstration of the boat's potential. *Brave Borderer* and *Brave Swordsman* were commissioned, although more economies and a concentration on larger frigates meant that no more Braves were built for the Royal Navy. *Brave Borderer* once sustained forepeak damage when being driven hard in a seaway, but the two boats were strengthened forward and gave no further trouble during their ten-year life in the Trials and Special Service Squadron. During one series of exercises, *Brave Borderer* maintained an average of 42 knots against a Force 7 headwind. In August 1962 they were attached to the Fishery Protection Squadron for the surprise interception of poachers in coastal waters. One of their other tasks was to mark the rolling start of the annual Cowes to Torquay race, and then accompany the powerboats to rescue the crews of any in difficulties.

The Braves immediately attracted the attention of foreign navies, encouraged by Vosper's private demonstration boat *Ferocity*. She was slightly smaller with a wooden hull 26·8m (88ft) long – 27·6m (89ft 6in) overall. *Ferocity* was more deeply V-ed forward and could be a little difficult to handle at first, but modifications soon eliminated any problems. An enclosed bridge enabled her to continue operating while sealed off from atomic chemical, and bacteriological fallout, which could be cleared by water sprays. However, an open bridge was also retained; it had advantages for night operations.

The Proteus engines could run for 2000 hours between overhauls, but their fuel consumption (like all gas turbines) was high. So only two were installed in *Ferocity,* each being coupled to a simpler gearbox for V-drive and reduction gears only. Cruising at 12 knots, manoeuvring and going astern, was provided by two Mathway/Daimler (General Motors) marine diesels. Synchro-self-shifting clutches with special interlocks prevented their engagement if the boat was going ahead at more than 15 knots. *Ferocity* could be armed in various ways to suit the customer's individual requirements.

It had been expected that emergent navies would be more interested in the cheaper two-shaft layout, which was still capable of 50 knots. In fact, only one of these was sold, the 28·9m (95ft) *Pfeil* to Germany. The 30·1m (99ft) three-shaft Brave derivatives proved more popular, all (except the German *Strahl)* being built with cruising diesels on the wing shafts. Nine were built for Germany, Malaysia, Brunei, Libya, and Denmark, with four more being constructed in Copenhagen. *Søloven, Søridderen, Søbjornen, Søhesten, Søhunden,* and *Søulven* carried heated torpedo tubes to protect their torpedoes from freezing spray in the Baltic. Most of the others had variations in armament or equipment. The Malaysian *Perkasa, Pendehar, Handalan,* and *Gempita* had propellers with reduced pitch to allow for the loss of gas turbine power in the tropics.

But Vosper was not alone in the production of FPBs. Other companies and other countries had been developing traditional designs and introducing fresh concepts. With an S-boat tradition behind her, it is not surprising that West Germany's revitalised navy should include a force of FPBs. It was also to be expected that they would be diesel-powered and built by Lürssen at Vegesack.

In 1954, the firm had begun launching the *Pleijad*-class for the Royal Swedish Navy. They displaced 157·5-172·7 tonnes (155-170 tons) and were 45m (147·6ft) in length. They carried two 40mm guns and six 53·3cm (21in) torpedo tubes, plus rocket flare launchers. Three 2238·0kW (3000bhp) diesels gave them a speed of 37·5 knots. From this recent experience, evolved the Type 55 (later known as the Type 140) *Schnellboot* for the Federal German Navy.

They had steel frames and alloy bulkheads with mahogany planking. The hull was round-bilge for almost 30m (100ft) becoming hard-chine farther aft. Their dimensions were 42·5m × 7·2m × 2·4m (139·4ft × 23·4ft × 7·9ft) and they displaced 162·6-193·0 tonnes (160-190 tons). Four 20-cylinder Mercedes-Benz 2238·0kW (3000bhp) diesels produced a speed of 42 knots, the shaft brackets being of thin spring steel. If one fractured the vessel could still return safely, for the shaft would be kept in position by the mass-energy of the turning propeller. The armament comprised two 40mm/L70 Bofors and four 53·3cm (21in) torpedo tubes, although four mines could be carried instead of two of the tubes. The Type 141 had 16-cylinder diesels made by Maybach and all forty of both type were named after animals and birds, for example, *Jaguar, Albatros* and *Ozelot.*

In 1959 Vosper built the 50-knot *Ferocity* to consolidate design improvements. Their subsequent craft of this type were of glued timber construction and fitted with twin diesels for cruising. She was smaller and less powerful than the Brave class but almost as fast, and carried virtually the same armament. She was not bought by the Royal Navy but her performance led to numerous orders for variant FPBs from foreign governments. In this photo she is carrying torpedoes with dummy warheads. *Vosper*

The design had considerable export potential. Four simpler 35·6m (118·1ft) vessels with two 1790·4kW (2400bhp) diesels were built in Spain for the Chilean Navy as *Fresia, Guacolda, Quidora* and *Tegualda.* The ten boats of the *Flyvefisken-* and *Falken-*classes were constructed in Denmark under licence, while Lürssen sold similar craft to Ecuador, Turkey, and Saudi Arabia. A further six went to Indonesia, *Madjan Tutul* being sunk by Dutch warships off Borneo on 15 January 1962.

Italy was another country with a fast fighting boat tradition. Wartime Higgins and MS-boats were still serving in her navy when the much larger *Folgore* was launched on 21 January 1954. She was 39·5m (129ft 6in) long and displaced 162·6-193·0 tonnes (160-190 tons), her four 1865·0kW (2500bhp) diesels giving her a rapid acceleration to 38 knots. *Folgore*'s eventual armament comprised two 53·3cm (21in) torpedo tubes and two 40mm guns. *Lampo, Baleno, Frecchia* and *Saetta* followed as convertible torpedo gunboats. The centre one of their three propellers was driven by a gas turbine, with a Fiat diesel on each wing shaft.

Italian influence could be traced in the Swedish MTBs built by Kockums Mekaniska Verkstads Aktiebolag at Malmö after the Second World War. Their 22m (75ft 6in) hard-chine hulls were built of steel and powered by petrol engines. The *T 32-* and *T 42-*classes were succeeded by the Lürssen boats of 1954, in turn followed by the *Spica-*class. Designed and built in Sweden, their radar-controlled armament comprised six 53·3cm (21in) torpedoes and one 57mm gun, plus ten rocket flare launchers. Three Bristol Siddeley Proteus 1274 gas turbines developed a total of 9489·1kW (12720shp), driving *Spica, Sirius* and their sisters at 40 knots. Displacing 203·2-233·7 tonnes (200-230 tons), their dimensions were 41m × 7·1m × 1·6m (134ft 6in × 23ft 4in × 5ft 2·4in). Later refinements included the employment of wire-guided torpedoes.

Norway, too, was active in this field. When her old PT-boats began to wear out, Jan H. Linge of the Batservice Verft, Oslo, designed a suitable replacement known as *Nasty*. She was 24·5m (80ft 4in) long, similar in size and armament to the Elcos, and powered by two Napier Deltic turboblown diesels. These gave her a range of 450 or 600 miles at 40 and 25 knots respectively. *Nasty* was the prototype for the *Tjeld-*class, supplied to the Greek, Turkish, and American navies (as well as Norway from 1960 onwards). Some of their hulls were composed as a mahogany-fibreglass-mahogany sandwich. The Storm-class was larger and armed as a gunboat with one 76mm gun, one 40mm gun, and depth-charge throwers. The 36·5m (120ft) hull displaced 101·6-127·0 tonnes (100-125 tons), two Maybach MB872A diesels producing a total of 5371·2kW (7200bhp).

East Germany was another country who had produced her own design of an FPB. She had three types, each about 17m (55ft) in length. Type 1 was flush-decked, while Type 2 had a raised forecastle. Both were armed with two 53·3cm (21in) torpedo tubes firing over the stern, whereas Type 3 mounted three tubes aft.

Argentina, Cambodia, Ethiopia and Israel were among those nations who also possessed small numbers of FPBs. Many of those in service throughout the world by the late 1960s were still of Second World War vintage or derivation and had changed hands several times.

The largest postwar supplier of fast fighting boats was the Soviet Union, whose craft served in the navies of Albania, Algeria, Bulgaria, China, Cuba, East Germany, Egypt, Finland, India, Indonesia, Iraq, Nigeria, North Korea, North Vietnam, North Yemen, Poland, Rumania, Somalia, South Yemen, Syria and Yugoslavia. Counted in hundreds, these boats were about to introduce a new weapon to the light coastal forces scene.

The Russian P2- and P3-classes had stemmed from the wooden hard-chine Vospers, Elcos, and Higgins PT-boats supplied during the Second World War. The P4-class built from 1951 to 1958 measured 19·1m × 3·5m × 1·7m (62ft 8·4in × 4ft 8·4in × 1ft 6in), the overall length being 22m (72·2ft). Because their hulls were of aluminium alloy (similar to the wartime G5-class) they only displaced 22·4 tonnes (22 tons). They were armed with two 12·5mm or 15mm machine guns, and two 45·7cm (18in) torpedo tubes. Two 820·6kW (1100bhp) diesels gave an estimated speed of 50 knots. Parallel with them were the wooden-hulled P6-class of 67·1-76·2 tonnes (66-75 tons) displacement. Being larger – 25·7m (26m overall) × 6·1m × 1·8m (84ft 2·4in [85ft 3in overall] × 20ft × 6ft), they had room for four 895·2kW (1200bhp) diesels resulting in a speed of 43 knots and a range of 450 miles at 30 knots. Their armament comprised four 25mm and two 53·3cm (21in) torpedo tubes. The MO6-class was an anti-submarine variant with depth-charge throwers and racks instead of torpedo tubes.

Later submarine chasers had gas turbines, keeping them in line with their torpedo-carrying sisters of the P8 (hydrofoils) and P10-series which followed. The Osa (Wasp)-class introduced in 1959 was larger still, of some 167·6-203·2 tonnes (165-200 tons). Their dimensions were 39·2m × 7·7m × 1·8m (128·7ft × 25·1ft × 5·9ft), three 3170·5kW (4250bhp) diesels giving a range of 800 miles at 25 knots and a top speed of 32 knots. They were armed with four 30mm guns and four box-like fixed missile launchers carrying the 4·6m (15ft) long SSN-2A Styx missile. Launched by a rocket motor under the fuselage a liquid fuel rocket motor then drove the Styx at subsonic speed for 23 miles. Although fitted with search and fire control radars, the Osa-boats had to maintain a straight course for five minutes before firing to allow the missile control gyros to start functioning. This meant that they could not be fired satisfactorily in a sea of more than state 4. Once launched and approaching the target, the missile's own radar took over control for the last few miles, although even this could be jammed – if the target were suitably equipped and could detect the oncoming missile. The Osa II series was the same as the Osa I-class, but with cylindrical missile launchers, probably to take the 6·4m (21ft) SSN-11, a modified Styx with folding wings. Its range was increased and it flew at a height of 50m (150ft) but the control problems seem to have remained the same.

As a cheaper and quicker means of getting the Styx missiles into wider service, P6 hulls were adapted to carry two Styx missiles. These were known as Komar (Mosquito)-class boats, but their operational potential was limited to seas of state 3 and less. Perhaps for this reason, the next two classes were more conventionally armed. The Stenka-class had a similar hull to the Osas, but instead of missiles carried four 40·6cm (16in) anti-submarine torpedoes and two depth-charge throwers. The Shershen-class displaced 152·4-162·6 tonnes (150-160 tons) and measured 35·2m (38m overall) × 7·0m × 1·5m (115ft 6in [125ft overall] × 23ft 1·2in × 5ft). Their armament comprised four 30mm guns, four 53·3cm (21in) torpedo tubes and twelve depth-charges. Three diesels produced a total of 9698·0kW (13000bhp) providing a speed of 41 knots. Although surface-to-surface missiles had arrived in FPBs, their use was confined to calm waters. It seemed that their installation in such lively craft had been just a propaganda exercise, but as has happened so often in the history of high speed attack craft, dramatic events suddenly changed the whole situation.

13 MISSILES, HYDROFOILS AND HOVERCRAFT

Small craft certainly found a rôle in the Vietnam War, the beginning of hostilities being marked by reported encounters between North Vietnamese MTBs and the American destroyers *Maddox* and *Turner Joy* in the Gulf of Tonkin on 2 and 4 August 1964; but this was really the only open-sea confrontation between the rival surface forces. The North Vietnamese P 4-, P 6-, Komar,- Shanghai- and Swatow-class boats suffered many losses from the retaliatory air attacks that were immediately ordered, while the Americans and South Vietnamese then had to face problems of coastal and riverine operations. Early on *Nasty*-class FPBs were employed in landing raiding parties, covering minesweepers, intercepting junks, and hurrying supplies and support to threatened units, but they proved too vulnerable in that environment and *PTF 4, PTF 9, PTF 14, PTE 15, PTF 8* and *PTF 16* were lost between November 1965 and August 1966.

Accordingly, some five hundred 9·8m (32ft) fibreglass RPBs (river patrol boats) were built. They were much cheaper than FPBs and only displaced 8·1 tonnes (8 tons). They could operate in shallow water because two General Motors diesels drove pumps sucking in water via a screen in the hull bottom and forcing it out through steerable nozzles. Floating in a foot of water, the RPBs could do 25 knots.

Generally, though, armour and firepower were preferred to speed when it came to surviving an ambush up some narrow, twisting, and overgrown delta waterway. ASPBs (assault patrol boats) 15·2m (50ft) long were specially built of welded steel for this work, although others were converted from landing craft. They were virtually river monitors and sometimes designated as such, being occasionally nicknamed 'battleships'. They could carry combinations of 81mm mortars, 40mm grenade launchers, 20mm guns, 0·50in machine guns, 0·30in machine guns, flamethrowers, and powerful water cannon to wash out earth bunkers along the river banks. In some, springs supported the armament and engines to minimise the shock of mine explosions. They were capable of 14 knots and their low freeboard completely restricted them to rivers. Something faster and more seaworthy but still inexpensive was needed for the Coastal Surveillance Force.

The 50·1m (164ft 6in) *Asheville*-class of CODOG-powered patrol gunboats drew 2·9m (9ft 6in) of water and were too big for working really close inshore. The immediate answer was the diesel-engined propeller-driven Swift-class of all-metal 15·2m (50ft) patrol craft, a simple modification of the boats used to ferry personnel out to Caribbean oil rigs at 28 knots. Mortars, machine guns, and grenade launchers made up the armament of these boats. Offshore frigates used their radar to direct the Swifts, covered them with heavier guns and replenished them every thirty-six hours. In one year alone, the Coastal Surveillance Force inspected half a million vessels to see if they were carrying supplies for the Viet Cong. By the end of the Vietnam War, the United States Navy had designed a CPIC (coastal patrol and interdiction craft) armed with four 30mm machine guns. As she was intended for rougher seas and a variety of environments, she was naturally larger than the Swifts. Her overall dimensions were 30·3m × 5·5m × 1·8m (99ft 2·4in × 18ft × 6ft) on a full load displacement of 72·4 tonnes (71·25 tons). The waterjet propulsion of 45 knots was provided by three 1342·8kW (1800shp). Avco Lycoming TF-25 gas turbines, although two 111·9kW (150bhp) diesels powered outboard drive propellers for cruising and manoeuvring.

It is interesting that all these specialised warships produced in response to Vietnamese riverine and estuarial warfare, were quite similar in appearance and armament to those employed by Soviet forces in their lake and waterway campaigns of the Second World War.

The Arab-Israeli conflict of 1967 saw some of the most significant developments of fast fighting boat warfare. Besides destroyers dating from the Second World War, both sides had FPBs. The Israeli Navy included *Daya*-class MTBs built by the French firm of Meulan between 1950 and 1956, together with the 21·3m (70ft) Baglietto *Ophirs* built between 1956 and 1957. Komars and Osas were numbered among their opponent's forces, but there was hardly any action during the Six-Day War itself. The Egyptian fleet succeeded in extricating itself from the dominance of Israeli aircraft over Port Said, and withdrew to Alexandria. But then the Israeli Navy could patrol the Egyptian coast practically unchallenged during the months of enmity that followed. On 11/12 July 1967 two Egyptian P6 FPBs intercepted two Israeli MTBs off Romani Bay. The latter were supported by the three 4·5in and six 40mm guns of the destroyer *Eilat*, who sank the two Egyptians at a cost of eight Israeli casualties.

On 21 October 1967 *Eilat* was still engaged upon her normal patrol work, which had now become a regular routine; so regular in fact, that her movements could be anticipated and tracked on radar. At 5.30pm *Eilat* was struck by two Styx missiles launched from a Komar-class boat within Port Said harbour. Listing heavily, she could not be saved, but the destroyer was still afloat at 7.30pm when two more Styx arrived. One landed nearby causing further casualties among the survivors in the water, while the other struck and sank *Eilat*. Forty-seven men were killed and ninety-one wounded out of a complement of 202. The Israeli reaction included air strikes on the missile craft's bases, but these were soon precluded by the arrival of neutral Soviet warships.

The Egyptians claimed that *Eilat* was 10 miles from their coast and thus well within their 12-mile limit. The Israeli Navy maintained that *Eilat* was at least a mile outside Egyptian territorial waters. What was important for the world's navies, was that the Styx missile had come from a Komar-boat.

Admittedly, these missiles had not been launched in the open sea. But it was now realised that the calm waters in the lee of any cove or jetty could conceal a tiny antagonist with a long-range punch as powerful as anything carried by larger surface warships. Israel, the first victim, was also one of the first to respond to this form of maritime warfare. She had already considered the acquisition of modern attack craft and *Eilat*'s loss confirmed this decision. Indeed one of the twelve new ships was called *Eilat*. Designed by Lürssen and

One of the last of Lürssen's Jaguar-class FPBs, built in 1964. The post-war boats were displacement hulls of composite construction and similar form to the wartime boats, but with greater beam to accommodate a fourth main engine.

built by Chantier Mécanique de Normandie, Cherbourg, the *Saar*-class was completed in 1969. Their overall dimensions were 45m × 7m × 1·8m (147ft 7·2in × 23ft × 5ft 10·8in), with a displacement of 223·5-208·3 tonnes (220-250 tons). Four 2517·8kW (3375bhp) diesels gave a speed of 40 knots, the endurance being 1000 miles at 30 knots and a phenomenal 2500 miles at 15 knots. They needed that endurance right at the beginning of their service when they escaped a trading embargo forbidding their delivery and made a secret passage to Haifa. They were originally armed with three 40mm guns or one 76mm Oto Melara gun, which required two loaders but was otherwise fully automatic. There were also two twin tubes aft, firing 53·3cm (21in) anti-submarine or surface torpedoes. On arrival in Israel the *Saars* were converted to carry Gabriel missiles, controlled by radar and optical sights, with heat-seeking and radar-homing devices. Launched from either fixed or trainable mountings, the Gabriel had a range of 11-20 miles at low altitude and its warhead weighed 154kg (338lb).

Soon the Styx missiles were giving further proof of their capabilities. On 13 May 1970 a 71·1-tonne (70-ton) Israeli fishing vessel was attacked and sunk at sea by Egyptian missile craft. Meanwhile, Osas were being supplied to the Indian Navy and several of them formed part of a task force which attacked Pakistan shipping off Karachi on the night of 4/5 December 1971. The Battle-class destroyer *Khaibar* was sunk and her sister *Badr* damaged by the Osas. Further hits were obtained on the coastal minesweeper *Muhafiz*. The boats returned on the night of 8/9 December 1971. This time their targets were merchantmen. A Panamanian and a British freighter were sunk, a Pakistan naval tanker damaged, and further damage done to shore installations.

Immunity was not always guaranteed to fast attack craft. Two Egyptian MTBs were destroyed by Israeli frogmen in Ras as-Sadat near Suez on 7 September 1969, while on 22 October 1970 Israeli aircraft accounted for two MTBs south of Shadwan Island in the Red Sea. Further air strikes dealt with a destroyer and a Komar off Ras Bandas on 16 May 1970 and in October 1973 several missile boats, merchantmen, and other vessels were destroyed or damaged by Israeli warships, themselves using Gabriel missiles.

Other navies were now taking serious note of the operational success of missile armed fast attack craft. Many adapted their own boats as missile carriers, or started to develop existing designs. By 1970 Norway's *Storms* were receiving Penguin surface-to-surface inertial, then heat-seeking, missiles with a range of 11 miles and a warhead of 1189kg (264lb). The *Snogg*-class used the same hull to carry a formidable array of four Penguin missiles, four 53·3cm (21in) torpedo tubes and one 40mm gun. *Jägaren* built for the Swedish Navy was somewhat similar. Among Finland's own designs was *Isku*, a box-like hull 26m × 8·7m × 2m (86ft 6in × 28ft 7·2in × 6ft 7·2in) in size. Completed in 1970, *Isku* had a speed of 25 knots and carried four Styx missiles. West Germany's Lürssen Werft was responsible for the 47m (154ft 2·4in) Type 148 and the 57m (200ft) Type 143, displacing 237·7-269·2 tonnes (234-265 tons) and 299·7-384·0 tonnes (295-378 tons) respectively. They reverted to traditional *Schnellboot* numbering as *S 41* to *S 70* and were powered by four 2834·8kW (3800bhp) MTU diesels with a speed of 38 knots. The larger boats had a range of 1300 miles at 30 knots, compared with the others' 600 miles. Both mounted two twin launchers for the French Aerospatiale Exocet MM38 homing missile with a 90kg (200lb) warhead and a range of 20 miles. The Type 143 also had 53·3cm (21in) wire-guided torpedoes in two tubes right aft. The gun armament could comprise Oto Melara 76mm and Bofors 40mm weapons. The Type 143 was similar to the modified *La Combattante*-class built in France.

There has been considerable Franco-German cooperation in recent years in the design and construction of fast attack craft. The German 406.4-tonne (400-ton) Type 143 are the largest diesel-powered FACs ever constructed. Built by Lürssen and similar to CMN's *La Combattante* designs, they are equipped with Europe's most advanced surface-to-surface missile, the Exocet MM 38. This is the 57m (187ft) *S 61* seen in 1975.

P6111

Although not armed with missiles, the Japanese *PT 11, PT 12, PT 13, PT 14* and *PT 15* were built with Mitsubishi diesels and IHI gas turbines, giving them a speed of 40 knots. Meanwhile other countries were equipping their forces with patrol craft of 25 knots and less. Typical was the 37·5m (123ft) *Mamba,* built by Brooke Marine of Lowestoft for Kenya. Powered by two 16-cylinder Ruston diesels and displacing 127·0-162·6 tonnes (125-160 tons), she mounted two 40mm guns. The same firm supplied comparable vessels to New Zealand, Nigeria, and other navies. Associated British Tool Makers Ltd built the Mexican Azteca-class. They were 34·1m (118ft 10in) long with a speed of 20 knots and were hardly high speed attack craft, but substitute missiles for guns and such a design could become a very formidable opponent in coastal waters. So economical, reliable, and well-tried diesels in round-bilge hulls flattening out aft to allow a near-planing condition at speed were particularly attractive to smaller countries with limited funds. Several had been supplied by Vosper Thornycroft, who had merged in 1966. This firm had also fitted stabilisers to smaller craft, permitting a higher weapon operating speed.

Some Vosper Thornycroft fast attack craft for overseas buyers were now being equipped with guided missiles. The Libyan *Susa, Sirte* and *Sebha* were intended for operations from the FPB support ship *Zeltin,* similar in appearance to a landing ship dock. The three FPBs were armed with eight Nord-Aviation SS-12 guided missiles in two launchers at the side of the bridge. The aimer sat in a shock-absorbing seat, aligning his sight with the target and directing the missile via electrical impulses sent along a fine wire run out from behind the missile. The SS-12 had a limited range of 3 miles and was difficult to control, but its 30kg (66lb) warhead arrived with the impact of a 11·4cm (4·5in) shell. The Brave derivatives supplied to Malaysia and Brunei were retrospectively fitted with SS-12 launchers early in the 1960s.

Brooke Marine's *Mamba,* a 36-knot FPB for the Kenyan navy built in 1974, is a class developed from long-range support and recovery vessels designed for the Ministry of Defence.

Beam-riding and homing missiles had a greater effective range of 12 miles and more, but they did need a larger vessel for better launch stability. Although such a craft would be non-planing and thus slower, this would not matter so much because of the missile's greater range and better chance of hitting. Once again Vosper Thornycroft decided to meet a hypothetical requirement by constructing another private venture boat, embodying their latest ideas and experience. Laid down in January and launched on 18 February, *Tenacity* was completed on 24 September, all in 1969. She had a round-bilge steel hull, with a spray-deflecting chine forward and a flat run aft. The superstructure was aluminium and she displaced 167·6-223·5 tonnes (165-220 tons). Her dimensions were 44·1m × 8·1m × 2·2m (144 ft 6in × 26ft 8in × 7ft 9in). Her CODOG machinery comprised three Rolls-Royce 3170kW (4250shp) Proteus gas turbines, with two 6-cylinder Paxman Ventura diesels on the wing shafts. Their cruising speed was 16 knots, compared with the maximum speed of 40 knots. *Tenacity* was intended to take Swiss Contraves Sea Hunter fire control radar for guidance of both anti-aircraft gunfire and Sea Killer missiles, but this armament was never fitted although mock-up weapons were put on board.

Although the Ministry of Defence did not order any of this class, the Royal Navy was aware of the threat posed by fast attack craft. Practical training in defence against them had been provided by the Braves, but they were reaching the end of their life. Accordingly, the Ministry of Defence ordered three unarmed fast attack craft from Vosper Thornycroft. *Scimitar* was launched on 4 December 1969, followed by *Cutlass* and *Sabre* on 18 February 1970 and 21 April 1970 respectively. They had hard-chine planing hulls of glued laminated wood, measuring 31·5m × 8·3m × 1·5m (103ft 6in × 27ft 8in × 5ft), and displacing 103·6 tonnes (102 tons) full load. A CODOG arrangement on two

Vosper Thornycroft's *Tenacity* was built in 1969 to stimulate sales for a larger type of fast attack craft which was then being designed in other European shipyards, principally to support missile systems. She was leased to the Royal Navy in 1971 and in 1972 was bought outright for fishery protection duties. *Vosper*

HMS *Sabre* and HMS *Cutlass* at Vosper Thornycroft's yard in 1970. They were ordered by the Ministry of Defence to provide training experience against boats of the Russian Osa and Komar types. *Vosper*

shafts was powered by two 3170kW (4250shp) Rolls-Royce Proteus gas turbines (40 knots) and two Foden 242·5kW (325bhp) diesels (11½ knots). Their complement numbered twelve, compared with thirty in *Tenacity,* who later joined them for exercises with NATO ships and helicopters off Portland.

Some experts recommended fast attack craft for the defence of British oil rigs in the North Sea. However, the Ministry of Defence considered that an assault on these installations by vessels of this type unlikely, especially in the rough weather so common in those waters. It was decided that it was better to invest available money in more flexible frigates with all-weather capability, rather than in fast attack craft of limited range and seakeeping. Local support for the oil rigs could be provided by large motor patrol craft and by 16-knot trawlers equipped with helicopters.

All these modern designs are noteworthy for their use of new materials and their novel employment of traditional forms. Each medium has its advocates and its disadvantages. Glass reinforced plastic (GRP or fibreglass) provides a good strength/weight relationship, requiring little maintenance, but if badly knocked it can eventually weep because of its internal porosity. It can also creep and decompose at 250-300°C (482-572°F). In the tropics, GRP repair materials have to be kept in refrigerated stores. Once it is worn out, it cannot be melted down and used again. Wood is available all over the world and its properties are common knowledge. Modern adhesives, coatings and paints provide greater strength and longevity, but it can encounter prejudice because of its susceptibility to rot, fire and marine borers. It is also expensive, while the condition of the finished product depends on the skill of the individual craftsman, which does not endear it to a world of uniform standards. Ferrocement is quite heavy and is not widely used, although its list of advantages is impressive: everything can be cast; its strength increases with age; it is flexible and fire-resistent; it is not susceptible to marine borers, while barnacles and algae can be cleaned off with detergent (as indeed can be done to other hulls); it does not need painting. Steel, on the other hand, is very popular, corrosion and rust being controlled by coatings, paints and cathodic protection. It is very strong, can be welded, is watertight, and fire-resistent. It does interfere with magnetic compasses, but this phenomenon is well-known. At one time, aluminium with its extraordinary strength/weight relationship seemed the best possible material for light fast craft. But, although it is resistent to salt water, it can be corroded by fresh water and is adversely affected by other metals so that it has to be insulated from adjacent materials. Its low melting point of 625-660°C (1157-1220°F) can be an additional fire hazard.

Besides developments in materials, engines, and weapons, this most recent period has seen an interest in unorthodox types of fast attack craft. The spectacularly fast has always held an appeal, and hydrofoils are certainly spectacular and fast. Seagoing hydrofoils are of two main types; those with surface-piercing foils and those with fully-submerged foils. Surface-piercing foils are inherently stable for an increase in wetted area means an increase in lift and a natural correction of the roll or pitch which has caused the extra immersion, while the faster the boat goes the more lift is generated. Modern surface-piercing foils are fixed and usually of V-form encircling the lower part of the hull, unlike the original ladder type. The fully-submerged foil (lowered on a retractable strut) is less resistent and provides lift and suction like an aeroplane's wing in the air, but it has to be controlled by manual or automatic means all the time during take off, flight, and landing (again like an aeroplane).

A hydrofoil creates less friction so it is more economical in fuel at high speeds than a conventional boat. The hull is carried above the waves so the stability of a 203·2-tonne (200-ton) hydrofoil may match that of a 3048-tonne (3000-ton) displacement vessel up to sea state 5, with waves 3·0-5·0m (10-15ft) high. Beyond that the hydrofoil either has to contour the waves with possible consequential shocks to equipment and personnel, or its legs have to be lengthened inordinately to clear the mounting waves when it platforms. This in turn means a corresponding increase in power to get the boat up on the foils: which may become prohibitively excessive. In any case, a long and complicated power transmission and control mechanism is required. If water propellers are used the necessary high speed may result in considerable cavitation, the phenomenon affecting the foils themselves. Hydrofoils are very difficult to handle at low speed and when manoeuvring in a constricted space because of the width and drag of the foils. Of course, as mentioned earlier, they can be retracted, and none of these problems are impossibly difficult, but the practical solutions are usually complicated and costly, and the more complications, the greater the risk of breakdown. But speed is vital in military operations, especially during withdrawal after a stealthy attack, a point appreciated by a number of scientists and builders in the years of the Second World War. There was some interest shown in Britain but most research was being carried out in Germany where Dr Attenkirch was at work.

The ideas of Hanns von Schertel were embodied in several hydrofoils with stepped hulls built by the Gebruder Sachsenberg Shipyard at Dessau-Roslau. Experimental craft were followed by the steel-hulled *VS 6*. She was 15·7m (51ft 7·5in) long, and displaced 17·3 tonnes (17 tons). Her two 581·9kW (780bhp) Hispano-Suiza petrol engines gave her a speed of 47 knots, compared with the 30 knots achieved by a conventional minelayer of similar proportions. *TS 1-6* had a single Lorraine Dietrich 12EB petrol engine driving contra-rotating propellers. Armed with a turreted machine gun, *TS 1-6* was designed for coastal surveillance. Following on from the wooden *VS 7*, came *VS 8* – an 81·3-tonne (80-ton) tank transporter – and *VS 10*, a 46·7-tonne (46-ton) torpedo boat. Her aluminium hull was 23·3m (82ft 2·5in) long. Four 1119·0kW (1500bhp) Isotta Fraschini ASM petrol engines were arranged in pairs to drive two propellers. Apart from an army workboat, the other German wartime hydrofoil was a torpedo boat of just 8·5m (27ft 11in) length, displacing 3·0 tonnes (3 tons). Her engine was a single 537·1kW (720bhp) Avia (Hispano-Suiza) V36. Reaching a speed of 52 knots on the Berlin lakes, she was intended to creep up on her target, launch a torpedo over her stern, and escape at full speed. Accident, capture, air raid or complete disappearance befell these craft during the last days of the Second World War.

The Italian designed hydrofoil *Sparviero.*

Since then several countries have studied and acquired hydrofoil warships. *Camiguin* and *Siquijor* were built by Cantier Navale Leopoldo Rodriguez of Messina. They were 21m (67ft 9in) long and were powered by two Mercedes-Benz diesels, driving propellers and producing a foil-borne speed of 38 knots. Basically commercial ferries, they were equipped with one 20mm gun and used on anti-smuggling patrols by the Philippine Navy, who also purchased two similar craft from Japan.

A more recent Italian design on the lines of the American *Tucumcari,* is *Sparviero*. She displaces 63·5 tonnes (62·5 tons) and is powered by a 33·6kW (4500shp) Rolls-Royce Proteus gas turbine driving a waterjet. *Sparviero*'s maximum speed is 50 knots, but cruising at 42 knots in sea state 4 gives a range of something over 400 miles. An endurance of 1200 miles can be achieved when hull-borne at 8 knots and driven by a diesel engine and retractable propeller. *Sparviero* was built at La Spezia by the Oto Melara Company around their 76mm gun and two of their Otomat homing missiles with a range of 32 miles.

The German hydrofoil, *VS 7*, built in 1943; the design included provision for a gun mounting on the superstructure but a weapon was not fitted.

The Royal Canadian Navy has undertaken an intensive programme to examine the feasibility of an anti-submarine hydrofoil. Hull-borne cruising and search was considered as important as foil-borne speed. Many practical experiments eventually indicated a preference for the non-retractable, surface-piercing canard layout. The result was *Bras D'Or,* designed by De Havilland (Canada) Ltd, and with a maximum take-off displacement of 238·8 tonnes (235 tons). She measured 46m × 6·6m (151ft × 21ft 6in), although the foils spanned 20·1m (66ft). When hull-borne, *Bras D'Or* drew 7·2m (23ft 6in), but 2·3m (7ft 6in) when up on the foils. The all-welded aluminium hull was constructed by Marine Industries Ltd at Sorel (Quebec). The bow foil served as rudder, while the after points of the stern foils carried 2000rpm 1·22m (48in) supercavitating propellers driven by a 16412·0kW (22000shp) Pratt & Whitney FT4A-2 gas turbine. Halfway up the foil legs were two more propellers, but these were 2·12m (84in) in diameter and were reversible. They were driven at 315rpm by a 1492·0kW (2000bhp) Davey Paxman 16YJCM diesel. *Bras D'Or* was able to take-off at 12 knots and achieve a speed of 63 knots in 1·8m (6ft) seas, but she was never armed. When trials were concluded she was laid up at Halifax in 1971.

American postwar research favoured fully-submerged foils, much attention being paid to the development of autopilots to sense changes in foil depth and make the appropriate corrections. Gas turbines were also employed in a number of experimental craft including a conversion of the famous wartime amphibious truck, renamed *Flying Duck*. The first American hydrofoil warship was designed by the Bureau of Ships, built by Boeing and completed in 1963. The 129·0-tonne (127-ton) *High Point* had propellers on the hydrofoil legs which slid upwards into the hull. Lockheed's *Plainview* was also driven by propellers (made of titanium), but she displaced 333·2 tonnes (328 tons) and her legs swung upwards through 180 degrees. Similar type of retraction was used in Grumman's *Flagstaff* and Boeing's *Tucumcari,* who were much smaller at 68·1 and 65·0 tonnes (67 and 64 tons) respectively, and were driven by waterjets. *Plainview* and *Flagstaff* had two foils forward and one aft, *High Point* and *Tucumcari* the opposite layout. These vessels mounted a variety of armament, but were essentially experimental craft, although *Tucumcari* did serve for a while with the Coastal Surveillance Force off Vietnam. She was later wrecked beyond repair when she ran aground on 16 November 1972.

In 1969 NATO's Commander-in-Chief South had asked for a substantial number of light forces to patrol straits and deal with the growing threat from missile-armed fast attack craft in the Mediterranean. The low cost, but high speed and manoeuvrability of small vessels would have to be combined with the stability of a larger ship. It was eventually decided that the United States Navy's fully submerged foil design would be best, but soon only three nations (the United States of America, West Germany, and Italy) were continuing in the project, although others sent observers.

Pegasus was built by the Boeing Company, being laid down at Seattle on 10 May 1973. Embodying the results of earlier United States Navy experience, she was launched on 9 November 1974 and commissioned in May 1975. *Pegasus* has an all-welded aluminium alloy hull with a displacement of 224·5 tonnes (221 tons). Her overall dimensions with foils extended are 40m × 8·9m × 7·1m (131ft 2·4in × 29ft × 23ft 2·4in). The forward foil leg (which acts as a rudder) swings forward, and the two stern ones swing aft for stowing. In this position *Pegasus* measures 45m × 8·9m × 2·9m (147ft 6in × 29ft × 9ft 6in). The after foils are fitted with trailing edge flaps for control and contain water intakes feeding two pumps up in the hull. These are operated by a power-splitting reduction gear driven by a General Electric 13428kW (18000shp) gas turbine. For displacement operation (when stern rudders are brought into use), water is sucked up through a grille in the hull, each pump being powered by one 596·8kW (800bhp) Mercedes-Benz diesel. From the pumps, the water is forced through nozzles in the stern, giving *Pegasus* a foil-borne range of 500 miles at 40 knots in 2·5-4·0m (8-12ft) seas.

A single helm in the circular bridge directs the Automatic Control System which senses deviations to ascertain when the ship is ready for take-off, and to determine banking, trim, and altitude when foil-borne, and to control landing. The accommodation equipment includes a recycling and distillation system which can produce 160 litres (35 gallons) of water per man per day, maintaining a constant storage of 1137 litres (250 gallons) to avoid a fluctuating effect on the ship's performance. The complement comprises the captain, four officers, four petty officers, and fourteen ratings. The US Mark 92 (formerly the Dutch WM-28) fire control system directs the Oto Melara 76mm gun and two quadruple Harpoon launchers. These missiles are preprogrammed by radar and have their own radar homing devices. A range of 60 miles is claimed for them. *Pegasus* herself can operate for five days' patrol before replenishment. Thirty of these patrol combatant (missile) hydrofoils are planned.

Naturally enough, the Soviet Union is also taking an interest in hydrofoil warships, two main classes being in service. The 71·1-81·3-tonne (70-80-ton) Pchelas can do 50 knots, but are mainly used for coastal patrol by the KGB. The Turya-class of 1973 is much larger at 167·6 tonnes (165 tons). Measuring 37·5m × 8·5m × 1·8m (123ft × 27ft 10·8in × 5ft 10·8in), they carry two 57mm and two 25mm guns plus four 53·3cm (21in) torpedo tubes.

The largest number of hydrofoil fast attack craft has been produced by the Chinese. It has been estimated that seventy Hu Chwan-class boats have been completed in the Hutang Yard at Shanghai. Four light guns and two 53·3cm (21in) torpedo tubes form the armament on a 45·7-tonne (45-ton) metal hull, 21·3m × 5m × 1m (70ft × 16ft 6in × 3ft 1in). Diesel-powered, they can do 55 knots and have a range of 500 miles. Vessels of this class have been supplied to Albania, Pakistan and Rumania.

Although hydrofoil critics say that the type is so complex that it would be better to complete the dynamic sequence and build an aeroplane, some of their protagonists claim that the hydrofoil is practically unsinkable. On the other

The Royal Navy's 65-knot Wellington-(BH 7) class hovercraft (designed and built by the British Hovercraft Corporation) and Vosper Thornycroft's *VT 2* venture, which is capable of mounting four Exocet missiles, and twin 35mm Oerlikons, with a range of 600 miles at 43 knots. The large ducts at the stern reduce the noise of the air screws.
Vosper

hand, hovercraft advocates point out their design's ability to cross shallows sand, and swamps, ice, submerged defences, and minefields – all at 50 knots. There is no doubt that surface effect vehicles have a military potential, but so far they have been used as landing craft. Light machine guns give close-range support to assault troops and cargo handlers, but they have not been widely employed as well-armed seagoing warships.

An exception is the Imperial Iranian Navy, which has a squadron of British Hovercraft SRN 6 Winchesters and BH 7 Wellingtons. Some are employed as transports, but the Winchester Mark 4 and the Wellington Mark 5 have more than light machine gun armament, being used as fast attack craft for coastal defence and anti-smuggling patrol.

The Wellington Mark 5 weighs 45·7 tonnes (45 tons) and can carry Exocet missiles on her side decks, being 23·8m long and 13·9m wide (78ft 4in × 45ft 6in). A single Rolls-Royce/BS Marine Proteus 15M 541 gas turbine of 3170·5kW (4250shp) drives both the lift fan and the swivelling aerial propeller. The Wellington Mark 5 has a range of 500 miles at 65 knots over 1·0m (3ft) waves. Even though speed falls off drastically in rough weather, it is still 35 knots over 1·5m (5ft) waves.

There seems no reason why hovercraft should not be more widely employed as fast attack craft. Certainly there is no lack of design projects. Vosper Thornycroft have suggested a 101·6-tonne (100-ton) development of their VT-1. Two 1492kW (2000shp) gas turbines could power water propellers on skegs below a full peripheral skirt, giving a speed of 46 knots and a range of 600 miles. It should provide a stable seakeeping ability for an armament of Exocet missiles and twin 35mm guns directed by Contraves fire control.

In 1972 American firms completed two 101·6-tonne (100-ton) experimental surface effect ships, both about 25m (80ft) long and 11-13m (35-40ft) wide. They are built of aluminium with rigid sidewalls and both have a speed of 80 knots. Aerojet's *SES 100A* has four Avco-Lycoming TF35 gas turbines, their total 8952·0kW (12000shp) driving three fans for lift and two pumps for waterjet propulsion. *SES 100B* (built by Bell Aerospace) has a total of six engines. Three are United Aircraft of Canada 373kW (500shp) ST-6J-70 gas turbines driving eight lift fans. The other three are 3357kW (4500shp) Pratt & Whitney FT-12 gas turbines, and they power two supercavitating water propellers. One of these vessels has test-fired a missile vertically clear of the ship and wave clutter to home on to a target five miles away.

An even larger 2032-tonne (2000-ton) vessel has been proposed by America, while the Russians have suggested the Ekranoplan craft. A ground effect vehicle with a boat-shaped hull 121·9m (400ft) long, would be fitted with a 38m (125ft) wing and ten gas turbines. Not only would she have a speed of 300 knots but she could also be seaworthy in rough weather.

These are all projected, research, or developmental craft, testing the theories and speculations of their inventors. They may well be the fast attack craft of the future, but at present a hovercraft is generally regarded as 'it', rather than 'she'. Their greatest drawback is that they do not look like ships or boats. Until they do or until their potential as seagoing warships overcomes this prejudice, it is unlikely that they will appeal to the world's navies.

14 NEW TACTICS OR OLD STRATEGIES?

It has been suggested in the 1975-76 edition of *Jane's Fighting Ships* that 'a small power possessing fast attack craft now occupies a position where it can deny sea areas to the ships of far larger navies. The passage of straits and narrow waters ... could be totally inhibited by high speed, missile armed craft ... Difficult targets themselves, these craft are possessed of a power of destruction hitherto unbelievable'.

Certainly the armament of modern high speed attack craft is most impressive, but such claims have been made since the earliest steam torpedo boats. Yet at no time have fast fighting boats alone prevented the passage of a determined force of large, fast surface warships with all their defences manned and alert. Handicapped by some special mission, tied to a convoy, expecting some other form of attack, or no attack at all, then warships (about fifty from cruisers to submarines and trawlers of all nations) fell victim to MTBs during the Second World War. Not all these were destroyed outright on the high seas. Some were crippled and had to be scuttled, while others were wrecked in harbour and settled on the bottom. Attacking a fleet safe in a defended port is the other traditional employment of fast attack craft. Indeed, it worked against an unsuspecting foe, but once he came to expect this form of warfare, little was achieved by repeated assaults, even when the anchorage was a vast area like the Normandy beaches.

However, warships rarely force their way into disputed waters just to prove a point. This may happen occasionally in peacetime, and if they then suffer damage and casualties it does prove something. On the other hand, it is frequently necessary for mercantile and military transports to pass through a hazardous zone during hostilities, either to ensure their country's survival or to forward its policies. It is vital that these ships suffer neither damage nor casualty, yet it is these vessels that suffered most from fast attack craft during the Second World War.

PT-boats and British MTBs sank 140 merchantmen totalling 165,252 tonnes (162,650 tons) in European and Mediterranean waters, together with countless barges and junks in the Far East. Meanwhile S-boats accounted for ninety-nine Allied merchant ships amounting to 233,350·8 tonnes (229,676 tons). Admittedly this latter figure only represents 1·1 per cent of the total Allied loss, compared with 6·5 per cent sunk by mines (including those laid by S-boats) and 68·1 per cent by U-boat attack.

Submarines were thus the greatest danger to an ocean-going merchant fleet during the Second World War. But at some time the merchantmen had to approach their destination, or some focal point, or were forced into shallow waters by submarines, or had to shift cargoes along the coast because overland communications were inadequate. Their movements were further restricted by daylight air attacks and by minefields in areas unsuitable for large, well-armed escorts. It was in these situations that fast attack craft were most feared and performed most effectively.

It is unlikely that these basic principles have changed, although fast fighting boats have weapons of 10 miles' range and more. Indeed, it is possible that the basic forms of defence have not changed either. Perhaps some clues to the most effective type of defence may be provided by the table of fast fighting boat losses which follows. It is of necessity an approximation, for many boats fell victim to several agents simultaneously. It includes constructive total losses, but not vessels scuttled in harbour, to avoid capture, or destroyed before completion. It excludes MLs, MMSs, patrol craft, and expendable vessels like the Italian *barchini.* The Italian figures omit those taken over by the Germans after 1943, but includes the vessels operated by Italians serving with both the Fascist and Allied forces. The German list, however, encompasses all captured craft right up to the end of the war.

The losses by accidental fire show the danger inherent in the use of petrol as a fuel. The twenty-three PT-boats wrecked reflect their principal employment in enemy waters where charts were marked with such encouraging comments as 'White water seen here by Captain Bligh but position doubtful'. The risk of accidental collision during high speed formation manoeuvres was always present, and so too was the greater possibility of mistaken attack when concentrated numbers of Allied MTBs increased towards the end of the war. This particular factor may be even more likely in the future, when homing missiles are involved. Although ships can be fitted with electronic identification it may be switched off so as not to reveal the vessel's presence to the enemy. The seaworthiness of these craft is demonstrated by the small percentage foundering in heavy weather, although fighting their weapons in such conditions was quite another matter. A higher number proved more vulnerable to shore artillery.

FAST FIGHTING BOAT LOSSES DURING SECOND WORLD WAR

Cause of Loss	Germany	Italy	UK	USA	Total	Per cent
	1939-1945	1940-1945	1939-1945	1941-1945		
Air raids in harbour	43	21	8	3	75	19
Surface ships	19	2	33	8	62	16
Mines	16	2	29	4	51	13
Wrecked aground	5	3	7	23	38	10
Accidental fire	1	2	23	6	32	8

FAST FIGHTING BOAT LOSSES DURING SECOND WORLD WAR

Air raids at sea	10	5	7	4	26	7
Accidental collision	7	2	10	3	22	6
Enemy fast attack craft	14	—	7	—	21	5
Shore batteries	4	1	4	6	15	4
Mistaken attack	1	—	4	7	12	3
Foundered at sea	—	1	5	—	6	2
Miscellaneous/unknown	15	1	11	2	29	7
Totals	135	40	148	66	389	100

It would seem that the only sure way to eliminate the menace from FACs is to destroy or occupy their bases. This was how the Hanseatic League defeated the Heligoland pirates in the Middle Ages and how the Barbary Corsairs were countered in the nineteenth century. Second World War air raids wrecked several S-boat harbours and it would appear that the modern equivalent would be a tactical nuclear strike. However, it is easy to imagine most situations when such an action would not be desirable, and even a conventional air attack would not be politically expedient.

Air raids at sea may not be so effective as one might expect, especially if the targets are operating by night with the possibility of mistaken identification. If Kamikazes are regarded as air-to-surface missiles, a fast attack craft's extreme manoeuvrability might just enable her to survive even modern air raids in the open sea, unless there were enough aircraft to swamp the boat's defences.

Apart from air raids in harbour, it would appear from the table that most MTBs were sunk while attacking warships or merchantmen and not by some sort of MGB. This is probably a restatement of the old idea of convoy as a means of offence. It is a waste of effort to scour the seas for the enemy; better to concentrate around a tempting target and wait for him. If there are enough warships available, they can ambush him along his route. If the enemy will not come out and the transports arrive undisturbed then that is the main purpose of sea power.

In the event of a hostile fast attack craft approaching, a general purpose frigate is probably the most effective antidote in the open sea. Just as an observed torpedo boat stood little chance against a well-armoured and heavily armed battleship, just as destroyers drove off MTBs, so frigates protected by their own electronic counter-measures can use their missiles against fast attack craft. No matter how sophisticated the latter becomes, the bigger ship can always carry just one more missile, or radar device, or be a little more stable, or have a little extra fuel in reserve. Her helicopter can strike far beyond the frigate's own eyes and weapons. Although fast attack craft may have a combat radius of 500 miles, it is unlikely that they could alone completely dominate the ocean beyond the reach of their own armament. Once they leave the obscurity of the coast, they are liable to be observed by all sorts of modern surveillance equipment, including that of their prey.

Of course, it is a different story if one or two small coasters and schooners (the latter perhaps of growing importance in a world short of energy) have to pass through narrow waters strewn with reefs, rocks and islets, any one of which can screen a hostile craft from radar and vision. Too dangerous for frigates, escort here can only be undertaken by protective boats, using their high speed and manoeuvrability to ward off attacks by ranging around and ahead of their charges. In such an engagement, rapid identification and communication would be as important as weapons. Guns, too, would still be as useful as missiles which have a slower rate of fire and which might home on to an adjacent cliff or a friendly vessel rounding the headland. With increasing miniaturisation, missile guidance packs could be issued to cargo ships for their own protection. Merchant service gunners on both sides contributed greatly to repelling MTB attacks during the Second World War. Nevertheless, mercantile armament was only a reinforcement, never a substitute for adequate escort. So small fast warships for escort and interception could well be necessary in the future.

It is, therefore, unwise to assume that only small powers need be interested in fast attack craft. Many small nations do not have the resources or industry to develop and construct these specialised hulls, engines, and armament, so they depend upon outside supply, which can be cut off when an emergency comes. The larger nation, herself often a supplier in peacetime, can during that period construct a few prototypes for her own navy, ready for mass-production if necessary and if time allows. Although expensive, they are obviously cheaper and quicker to build than the frigates which form a large proportion of the peacetime construction of an oceangoing navy. Of course, if an emergency does arise there may not be enough time for fast attack craft to be built. This is one of the risks of international politics and such a crisis may not be of the sort which fast fighting boats could do much about even if they existed. It would certainly not be advisable to have an armada of such craft in continuous service for they are costly to run and their crews could well feel disgruntled if reasons of peacetime economy prevent them being driven at their proper high speed. Attempts to use them as a 'fleet in being' would probably repeat the frustrations experienced by Second World War crews in the West Indies, the Aleutians and in India, who had all the related maintenance and operational problems without any definite results to show for their efforts.

However, it is probably a a good idea for every major fleet to have at least one group of fast attack craft in commission. Not only can they train with frigates in the defence against these boats, but they can provide seagoing and staff experience in the operation and administration of a complete squadron of FPBs. It is not easy to adjust one's thinking from cruising speeds of 10-20 knots over 5000 miles to 40-50 knots over 500 miles. Boats can exercise by evaluating aerial and coastwatchers' information, by proceeding at speed to an area, using a mobile land or seaborne base, or waiting alone for several days for targets of opportunity (which an aircraft cannot do), and then returning rapidly. This high passage speed would be the particular advantage to be derived from hydrofoils and hovercraft. These types may also be less vulnerable to mines, but this claim was also made for Second World War MTBs. It is not beyond mining ingenuity to devise something to catch any sort of craft.

Probably the greatest lesson of the Second World War is that prewar speculations are not fulfilled. Anything can happen. Who, in 1939, would have expected MTBs to be torpedoing destroyers in the Gulf of Riga, blowing up installations in Norway, laying mines off Falmouth, capturing schooners in the

Adriatic, landing commandos in the Black Sea, examining canoes in the Solomons or patrolling in the Aleutians? Policies, governments and situations can change so quickly, and no admiralty can say with absolute certainty that in ten years' time it will not be interested in the Alaskan fisheries, the Malagasy coasting trade or North Sea oil rigs.

One thing does seem likely. Whether of displacement, planing, hydrofoil, or surface effect configuration, the trend for fast attack craft to become larger will continue. The miniaturisation of fire control equipment usually means the installation of more, not fewer, systems, together with their associated back-up arrangements, fail-safe devices and indicator panels. More sophisticated and compact missiles can allow for reloading from the magazine. Electronic equipment must be carried to detect and jam the enemy's radar beam-riding missiles, but if self-contained, they can be fired from his navigational radar and switch on their own homing system near the target. Travelling at the speed of sound just above the waves, they may not be detected until just before impact. Warning radar and computerised information control may have just enough time to launch chaff (or window) dispensers and flares to decoy radar-controlled and heat-seeking missiles, or very high rate of fire guns can go into action, their shells equipped with proximity fuses which work at very low level and detonate within half a yard of a missile.

All this gadgetry requires a bigger hull, itself providing a more stable weapon platform. Special movement-damping seats (called crew suspension stations) could be adapted for missiles in planing boats, but missile launchers are not usually happy with the small craft hammering that human crews are prepared to accept for various human reasons.

Bigger vessels will probably be slightly slower, but that is offset by the missile's greater range, although there is still the question of relative speed between target and assailant, and passage speed to and from patrol area or ambush point. So there comes into being fast attack craft as large as Israel's *Reshef, Keshet, Romah* and *Kidon*. They are 58m (190ft 7·2in) long and displace 421·6 tonnes (415 tons), with a maximum speed of 32 knots and a range of 1500 miles. A complement of forty-five runs the ship and mans seven Gabriel missiles, launchers, two 76mm Oto Melara guns, two machine guns and four depth-charges.

The same Israeli Navy has recently ordered 6·1-tonne (6-ton), 8·5m (28ft) Firefish boats from Sandaire at San Diego. Built of fibreglass, they are powered by two 160·4kW (215bhp) Mercruiser V8 engines, giving a speed of 50 knots over a distance of 150 miles. They can be radio-controlled or carry a crew of five to operate a small surface-to-surface guided missile. A swarm of such mosquito craft, dashing in at high speed from all directions, could perhaps overwhelm the most sophisticated frigate afloat.

Perhaps.

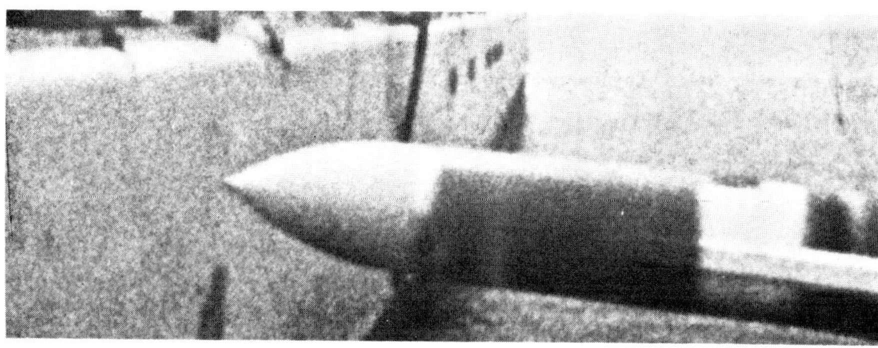

An Exocet *MM 38* at Mach 0.9, one-thousandth of a second before striking a stationary target ship.

NOTES

Introduction

1 Barker, *Small Combatants,* 1972.

Chapter one

1 Quoted fully in *Le Navigation Sous Marine,* G. L. Pesce (Paris 1906); also Archives Nationales, Dossier Marine D' 21, fol. 45-58.

2 Quoted in Dickinson, *Robert Fulton.*

3 Robert Fulton, *Torpedo War and Submarine Explosions.*

4 Quoted in Dickinson, *Robert Fulton.*

5 Quoted in Barnes, *Submarine Warfare.*

6 Commander W. B. Rowbotham, United States Naval Institute, *Proceedings,* December 1936.

7 *Niles' Register,* vol. IV.

8 Paper read in January 1867 to the Institution of Civil Engineers, *Transactions,* vol. 26.

9 E. J. Reed, *Our Iron-Clad Ships,* 1869.

10 Reply during the thirteenth session of the Institution of Naval Architects, 21 March 1872, *Transactions,* vol. 13.

11 Paper by N. Barnaby read to the Institution of Naval Architects, 6 April 1876, *Transactions,* vol. 17.

12 *A Lecture on the Whitehead Torpedo* at the USN Torpedo Station, 20 November 1874. (Quoted in Sleeman).

13 *A Short Treatise on Electricity and the Management of Electric Torpedoes,* 1868. *A Treatise on Electricity and the Construction and Management of Electrical and Mechanical Torpedoes,* 1871. *Addenda to the Second Edition of the Treatise on Electricity and Torpedoes,* 1873.

14 Public Record Office (PRO) Adm. 116/135.

Chapter two

1 Bramwell, 'On Quick Steam Launches', read to Institution of Naval Architects, 2 March 1872, *Transactions* vol. 13. Also *Engineering,* 3 November 1871.

2 C. M. Ramus, *The Polysphenic Ship.*

3 Letter to Admiralty 10 April 1972, PRO Adm 116/167.

4 Quoted in Phillips-Birtt, *Naval Architecture of Small Craft.*

5 Patent Office, 1877/418.

6 Quoted by John Donaldson in a lecture to the Royal United Services Institute, May 1877, *Journal,* vol. 21.

7 John Thornycroft, 'On Torpedo Boats and Light Yachts for High Speed Steam Navigation', paper read to the Institution of Civil Engineers 10 May 1881. *Proceedings,* vol. 66.

8 *Torpedo Manual,* quoted in Gray, *The Devil's Device.*

9 PRO. Adm 116/244.

10 Reply to John Donaldson's paper, 'The Further Development of the Thornycroft Torpedo Vessels', read 29 April 1881, to the Royal United Services Institute, *Journal,* vol. 25.

11 Armstrong, pp. 186-8.

12 J. A. Normand, 'On Sea-Going Torpedo Boats', paper read to Institution of Naval Architects, 15 March 1883, *Transactions,* vol. 24.

13 Admiral P. Colomb, 'The Future of the Torpedo', paper read to the Royal United Services Institute, 8 June 1897, *Journal,* vol. 41 ii.

14 Ibid.

15 *Turbinia* is now preserved in the Museum of Science and Engineering, Newcastle.

16 J. A. Normand's paper 'Note sur l'influence de l'immersion de l'hélice et de la vitesse sur la rupture du cylindre d'eau actionné', read to Association Technique Maritime, 1893.

Chapter three

1 Arthur Evans, 'The Internal Combustion Engine as a Means of Propelling Small Vessels', paper read to Institution of Naval Architects, 24 March 1904, *Transactions,* vol. 46.
2 J. A. Smith, 'The Design and Construction of High Speed Motor Boats', paper read to Institution of Naval Architects, 5 April 1906, *Transactions,* vol. 48.
3 *The Motor Boat,* 9 March 1905.
4 *Aeromarine Origins.*
5 *Fighting Ships,* 1905-6, edited by Fred T. Jane.
6 *The Motor Boat,* 16 March 1905.
7 *Fighting Ships,* 1905-6, edited by Fred T. Jane.

Chapter four

1 PRO. Adm 8372/74.
2 Ibid.
3 PRO. Adm 116/1479.
4 Admiral Sir Barry Domville, *By and Large.*
5 PRO. Adm 137/2273.
6 *MAS 15* is now preserved at the Museo del Vittoriano in Rome.

Chapter five

1 *CMB 4* was embalmed at Hampton for a long time but is now the property of the Imperial War Museum and is in the reserve collection at Duxford.
2 PRO. Adm 1/8571/298.
3 Harold Penrose, *British Aviation: The Adventuring Years 1920-1929.*
4 Quoted by Lieutenant Commander Henry H. Hower, *The Inside Story of the Harmsworth Trophy Contest of 1931.* (Pamphlet, Buffalo, NY, 1931.)
5 *Miss Britain III* is now preserved at the National Maritime Museum, Greenwich.
6 PRO. Adm 1 8771/165.
7 Ibid.
8 PRO. Air 2/1484.

Chapter six

1 PRO Adm 1 8771/165.
2 Patent no. 2879/35.
3 Patent Specification no. 478, 695, 18 January 1938.
4 *Parliamentary Debates,* 333 H.C. Deb. 5 s.
5 Ibid, 334 H.C. Deb. 5 s.
6 Ibid, 345 H.C. Deb. 5 s.
7 PRO. Adm 116/4043.
8 This boat has been renovated and is now used as a sea-going training vessel by Blofield and Brundall Sea Scouts, Norfolk.

Chapter seven

1 PRO. Adm 116/5316.
2 PRO. Adm 116/4736.
3 PRO. Adm 1/13896.
4 Admiralty, *Instructions for Coastal Forces Warfare,* London 1944.
5 Ibid.
6 Ibid.
7 W. J. Holt RCNC, 'Coastal Force Design', paper read to Institution of Naval Architects, 27 March 1947, *Transactions,* vol. 89.
8 PRO. Adm 1/17861.
9 Captain Augustus Agar, *Footprints in the Sea.*
10 PRO. Adm 116/4736.

Chapter eight

1 PRO. Adm 116/4043.

Chapter eleven

1 Captain Tameichi Hara, Fred Saito, and Roger Pineau, *Japanese Destroyer Captain.*
2 Dudley Pope, *Flag 4.*
3 Ibid.

SPEED CONVERSION TABLE

KNOTS		MPH
0·8	1	1·1
1·7	2	2·3
2·6	3	3·4
3·4	4	4·6
4·3	5	5·7
5·2	6	6·9
6	7	8
6·9	8	9·2
7·8	9	10·3
8·6	10	11·5
13	15	17·2
17·3	20	23
21·7	25	28·7
26	30	34·5
30·4	35	40·2
34·7	40	46
39·1	45	51·7
43·4	50	57·5
52·1	60	69
60·8	70	80·5
69·5	80	92
78·2	90	103·5
86·9	100	115·1

BIBLIOGRAPHY

Admiralty *Torpedo Drill Book,* HMSO, London 1908.

Admiralty 'Coastal Forces in the War: a historical survey' Public Record Office Adm 1/18997.

Admiralty 'Coastal Forces and HM Ships – Operations & Losses' Public Record Office Adm 199/541.

Admiralty 'War Diary Note Command Sept 1939-Dec 1940' Public Record Office Adm 199/375.

Admiralty 'Recoilless Guns for Coastal Forces Craft 1943' Public Record Office Adm 1/15376.

Admiralty *Ships of the Royal Navy – Statement of Losses During the Second World War* HMSO, London 1947.

Admiralty *British Merchant Vessels Lost or Damaged by Enemy Action During the Second World War* HMSO, London 1947.

Agar, Commander Augustus VC DSO 'Naval operations in the Baltic' Royal United Services Institute February 1928, *Journal,* Vol. 73.

Agar, Captain Augustus VC DSO *Footprints in the Sea* Evans Bros Ltd, London 1959.

Agar, Captain Augustus VC DSO *Baltic Episode* Hodder & Stoughton, London 1963.

Armstrong, Lieutenant G. E. *Torpedoes and Torpedo-vessels* George Bell & Son, London (2nd ed.) 1901.

Armstrong, Warren *HM Small Ships* Frederick Muller, London 1958.

Auphan, Rear-Admiral & Mordal, Jacques *The French Navy in World War II* (US) Naval Institute Press, Annapolis, 1959.

Bagnasco, S. Tenente di Vascello Erminio *I MAS e le Motosiluranti Italiane 1906-1966* Ufficio Storico della Marina Militare, Rome 1967.

Banbury, Philip *Shipbuilders of the Thames and Medway* David & Charles, Newton Abbot 1971.

Barker, Arthur Davidson 111 'Small Combatants – Communist, Neutral and NATO North' US Naval Institute (Naval Review May 1972) *Proceedings* vol. 98.

Barker, Arthur Davidson 111 'Small Combatants – The Oceanic Powers' US Naval Institute (Naval Review May 1973) *Proceedings* vol. 99.

Barker, A. J. *Suicide Weapon* Ballantine, New York 1971.

Barnaby, K. C. OBE *One Hundred Years of Specialised Shipbuilding and Engineering* Hutchinson & Co Ltd, London 1964.

Barnes, Eleanor *Alfred Yarrow: his life and work* Edward Arnold, London 1923.

Barnes, Lieutenant Commander J. S. USN *Submarine Warfare* Van Nostrand, New York 1869.

Bekker, C. D. *K-Men* William Kimber, London 1955.

Berger, Rear-Admiral P. E. C. MVO DSC 'The Royal Navy – A Concept of Maritime Operations', Royal United Services Institute 1974, *Journal* vol. 119.

Blackman, Raymond V. B. MBE *Ships of the Royal Navy* Macdonald & Jane's, London 1973.

Borthwick, Alastair *Yarrow: the first hundred years* Yarrow & Co Ltd, Glasgow 1965.

Bragadin, Commander (R) Marc' Antonio *The Italian Navy in World War II* Naval Institute Press, Annapolis 1957.

Brassey, Sir Thomas *The British Navy,* vol. II Longmans, Green & Co, London 1882.

Brice, Martin H. *The Tribals* Ian Allan Ltd, Shepperton-on-Thames 1971.

Brock, Lieutenant Commander P. W. 'Changes in naval warfare owing to new and modified weapons' Royal United Services Institute 1936, *Journal* vol. 81.

Brooke, G. A. G. DSC 'Navigational Aids for Small Craft' Royal Institution of Naval Architects, *Symposium on small craft,* 1971.

Buchard, H. *Torpilles et Torpilleurs* Berger-Levrault (Paris) & Libraires-Editeurs (Nancy) 1889.

Bulkeley, Captain Robert J. Jr. *At Close Quarters* United States Naval History Division, Washington 1962.

Cagle, Commander Malcolm W. & Manson, Commander Frank A. *The Sea War in Korea* Naval Institute Press, Annapolis 1957.

Chatterton, E. Keble *The Auxiliary Patrol* Sidgwick & Jackson, London 1923.

Chatterton, E. Keble *The Brotherhood of the Sea* Longmans, London 1927.

Chatterton, E. Keble *Seas of Adventure* Hurst & Blackett, London 1936.

Chun, Victor *American PT Boats in World War II* Private publication, Los Angeles 1976.

Clark, Robert 'Steel as a Boat Building Material' Royal Institution of Naval Architects, *Symposium on small craft,* 1971.

Clowes, W. Laird 'The place and uses of torpedo boats in war', Royal United Services Institute 1892, *Journal* vol. 36.

Cobb, David *HM MTB Vosper 70-Foot* Profile Publications, Windsor, 1971.

Colledge, J. J. *Ships of the Royal Navy* David & Charles, Newton Abbot 1970.

Colomb, Vice-Admiral P. H. 'The Future of the Torpedo' Royal United Services Institute 1897, *Journal* vol. 41 ii.

Cooper, Bryan *The Battle of the Torpedo Boats* Macdonald, London 1970.

Cooper, Bryan *The E-Boat Threat* Macdonald & Jane's, London 1976.

Couhat, Jean Labayle *French Warships of World War II* Ian Allan Ltd, Shepperton-on-Thames 1971.

Couhat, Jean Labayle *French Warships of World War I* Ian Allan Ltd, Shepperton-on-Thames 1974.

Cowie, Captain J. S. CBE *Mines, Minelaying and Minesweeping* Oxford University Press, London 1949.

Crane, Clinton W. *Clinton Crane's Yachting Memories* Van Nostrand, New York 1952.

Dawson, Christopher *A Quest for Speed at Sea* Hutchinson & Co Ltd, London 1972.

Dawson, Commander W. RN 'Offensive Torpedo Warfare', Royal United Services Institute 1871, *Journal* vol. 15.

Detmers, Theodor *The Raider 'Kormoran'* William Kimber, London 1959.

Dickens, Captain Peter DSO MBE DSC 'Narrow Waters in War', Royal United Services Institute 1969, *Journal* vol. 114.

Dickens, Captain Peter DSO MBE DSC *Night Action* Peter Davies Ltd, London 1974.

Dickinson, H. W. *Robert Fulton* The Bodley Head, London 1913.

Divine, David, *The Nine Days of Dunkirk* Faber & Faber, London 1949.

Domville, Admiral Sir Barry *By and Large* Hutchinson & Co Ltd, London 1936.

Donaldson, John M. Inst C.E. 'The more recent improvements in the Thornycroft torpedo boats', Royal United Services Institute 1889, *Journal,* vol. 33.

Donovan, Robert J. *The Wartime Adventures of President John F. Kennedy* Anthony Gibbs & Phillips, London 1961.

Dreyer, Rear-Admiral C. W. S. 'The Future for Fast Patrol Boats in the Royal Navy', April 1972, *Naval Review,* vol. 60.

Du Cane, Commander Peter CBE *High Speed Small Craft* Temple Press Ltd, London 1951, 1956, 1964; David & Charles, Newton Abbot (4th ed.) 1974

Du Cane, Commander Peter CBE *A History of the Principal Activities of Vosper during the Period 1931-1969* Private circulation, 1969.

Du Cane, Commander Peter CBE *An Engineer of Sorts* Nautical Publishing Company, Lymington 1971.

Du Cane, Commander Peter CBE 'Fast Patrol Boats', Royal Institution of Naval Architects, *Symposium on small craft* 1971.

Dunsany, Admiral Rt Hon Lord 'On the Laws and Customs of War as Limiting the Use of Fire-Ships, Explosive Vessels, Torpedoes, and Submarine Mines', Royal United Services Institute 1879, *Journal* vol. 22.

Eames, M. C. & Jones, E. A. 'HMCS *Bras D'Or* – An Open Ocean Hydrofoil Ship', read to Royal Institution of Naval Architects April 1970, *Journal,* April 1971.

Fitzgerald, Captain Fergus N. 'Ireland's Naval Policy', Royal United Services Institute 1948, *Journal* vol. 93.

Fock, Harald *Schnellboote* Koehlers Verlagsgesellschaft, Herford 1973.

Fock, Harald *Fast Fighting Boats* (Vol. 1) (Trans. Barbara Webb of *Schnellboote)* Nautical Publishing Company, Lymington 1977.

Fox, Uffa *Seamanlike Sense in Powercraft* Peter Davies Ltd, London 1968.

Fraccaroli, Aldo *Italian Warships of World War II* Ian Allan Ltd, Shepperton-on-Thames 1968.

Fraccaroli, Aldo *Italian Warships of World War I* Ian Allan Ltd, Shepperton-on-Thames 1970.

Fulton, Robert *Torpedo War and Submarine Explosions* (Facsimile of 1810 New York edition) Swallow Press, Chicago 1971.

Giddy, Commander O. C. H. DSC 'Light Coastal Forces in the Present War', Royal United Services Institution 1944, *Journal* vol. 89.

Granville, Wilfred & Kelly, Robin A. *Inshore Heroes* W. H. Allen, London 1961.

Gray, Edwyn *The Devil's Device* Seeley, Service & Co Ltd, London 1975.

Greger, René *Austro-Hungarian Warships & Naval Aircraft of World War I* Ian Allan Ltd, Shepperton-on-Thames 1976.

Grenfell, Captain H. RN 'The Position of the Torpedo in Naval Warfare', Royal United Services Institute 1889, *Journal* vol. 23.

Gröner, Erich *Die Deutschen Kriegsschiffe 1815-1945* J. F. Lehmanns, Munich, vol. 1. 1966, vol. 2. 1968.

Guest, Captain Freddie *Escape from the Bloodied Sun* Jarrold, London 1956.

Guns (pseud) *The Autobiography of a Whitehead Torpedo, Engineering* (London) & Griffin & Co (Portsmouth) 1886.

Hara, Captain Tameichi; Saito, Fred & Pineau, Roger *Japanese Destroyer Captain* Ballantine, New York 1961.

Harvey, Commander F. *Harvey's Sea Torpedo* E. & F.N. Spon, London 1871.

Herreshoff, L. Francis *Capt Nat Herreshoff* Sheridan House, New York, 1974 ed.

Hichens, Lieutenant Commander Robert Peverell DSO DSC *We fought them in Gunboats* Michael Joseph Ltd, London 1944.

Holden, Larry 'PHM – Patrol Hydrofoil Missile Ship', Royal United Services Institute 1974, *Journal* vol. 119.

Holman, Gordon *The Little Ships* Hodder & Stoughton, London 1943.

Holman, Gordon *The King's Cruisers* Hodder & Stoughton, London 1947.

Holt, W. J. 'The development of high speed craft for the Navy', Royal United Services Institute 1938, *Journal* vol. 83.

Holt, W. J. 'Coastal Force Design', Institution of Naval Architects March 1947, *Transactions* vol. 89.

Hope, Linton 'The effect of beam on the speed of hydroplanes', Institution of Naval Architects 1915, *Transactions* vol. 57.

Hummelchen, Dr G. *German Schnellboot* Profile Publications, Windsor 1973.

Kimche, David & Bawly, Dan *The Sandstorm* Secker & Warburg, London 1968.

King, Cecil *Atlantic Charter* Studio Publications, London 1943.

King, H. F. MBE *Aeromarine Origins* Putnam & Co Ltd, London 1966.

Kirby, G. J. 'A history of the torpedo', *Journal of Naval Science,* January and March 1972, vol. 27 i and ii.

Knight, E. F. *The Harwich Naval Force* Hodder & Stoughton, London 1919.

Landström, Björn *The Ship* Allen & Unwin, London 1961.

Leckie, Robert *The Korean War* Barrie & Rockliff, London 1963.

Le Fleming, H. M. *Warships of World War I* Ian Allan Ltd, Shepperton-on-Thames 1970.

Le Masson, Henri *Les Levriers de la Mer* Horizons de France, Paris 1948.

Le Masson, Henri *Histoire du Torpilleur en France* L'Academie de Marine, Paris 1963.

Le Masson, Henri *Guérilla Sur Mer* Éditions France-Empire, Paris 1973.

Lenton, H. T. *British Warships* Ian Allan Ltd, Shepperton-on-Thames.

Lenton, H. T. *German Surface Warships* Macdonald, London 1965.

Lenton, H. T. *Royal Netherlands Navy* Macdonald, London 1967.

Lenton, H. T. *Warships of the British & Commonwealth Navies* Ian Allan Ltd, Shepperton-on-Thames 1969.

Lenton, H. T. *German Warships of the Second World War* Macdonald & Jane's, London 1975.

Lenton, H. T. & Colledge, J. J. *Warships of World War II* Ian Allan Ltd, Shepperton-on-Thames 1964.

Lenton, H. T. & Colledge, J. J. *British Warship Losses of World War II* Ian Allan Ltd, Shepperton-on-Thames 1964.

Lund, Paul & Ludlam, Harry *Trawlers Go to War* Foulsham, London 1971.

Lupinacci, Capitano di Vascello Pier Filippo *La Marina Italiana nella Seconda Gerra Mondiale* Ufficio Storico Della Marina Militare, Rome 1961.

Macintyre, Captain Donald *Leyte Gulf – Armada in the Pacific* Macdonald, London 1970.

Mackay, Ruddock F. *Fisher of Kilverstone* Oxford University Press, Oxford 1973.

March, Edgar J. RINA *British Destroyers* Seeley, Service & Co Ltd, London 1966.

McKee, Alexander *The Coal-Scuttle Brigade* Souvenir Press, London 1957.

Meister, Jurg *The Soviet Navy* Macdonald 1972.

Monsarrat, Nicholas *The Ship that Died of Shame* Cassell, London 1964.

Moore, Captain John E. *The Soviet Navy Today* Macdonald & Jane's, London 1975.

Maxwell, Lieutenant Gordon S. RNVR *The Motor Launch Patrol* Dent & Sons, London 1920.

Normand, J. A. 'On sea-going torpedo boats', read to the Institution of Naval Architects March 1883, *Transactions* vol. 24.

Normand, J. A. *Etude sur les torpilleurs* Paris 1885.

North, A. J. D. *Royal Naval Coastal Forces* Almark Publishing Co Ltd, New Malden (Surrey) 1972.

O'Ballance, Edgar *The Electronic War in the Middle East 1968-70* Faber & Faber, London 1974.

Palit, Major-General D. K. VC *The Lightning Campaign* Thomson Press (India) 1972.

Pawle, Gerald *The Secret War* Harrap, London 1956.

Penrose, Harold *British Aviation: The adventuring years 1920-1929* Putnam & Co Ltd, London 1973.

Phillips-Birt, D. *The Naval Architecture of Small Craft* Hutchinson & Co Ltd, London 1957.

Poolman, Kenneth *The 'Kelly'* William Kimber, London 1954.

Pope, Dudley *Flag 4* William Kimber, London 1954.

Potter, John Dean *Fiasco* Heinemann, London 1970.

Pritchard, John *An Introduction to Royal Air Force high speed rescue launches 1938-48* Private publication, Royston, Hertfordshire, 1974.

Ramsey, Winston *The Italian Naval Attack on Grand Harbour* After the Battle, Plaistow, 1975.

Ramus, Charles Meade *The Polysphenic Ship* Edward Stanford, London 1874.

Ramus, Charles Meade *Rocket Floats and Rocket Rams* Edward Stanford, London 1875.

Ransome-Wallis, R. DSC MD *Two Red Stripes* Ian Allan Ltd, Shepperton-on-Thames 1973.

Redwood, Bernard R. 'Motor Boats', Society of Arts, March 1906, *Journal* vol. 54.

Reed, E. J. CB *Our Iron-Clad Ships* John Murray, London 1869.

Révue Maritime et Coloniale (Trans. Lieutenant E. Meryon RN) 'The Employment of Torpedo Boats against Ships', Royal United Services Institute 1880, *Journal* vol. 23.

Reynolds, Lieutenant L. C. DSC *Gunboat '658'* William Kimber, London 1955.

Rohwer, J. & Hummelchen, G. *Chronology of the War at Sea 1939-1945* Ian Allan Ltd, Shepperton-on-Thames 1974.

Roscoe, Theodore *United States Destroyer Operations in World War II* Naval Institute Press, Annapolis 1953.

Roskill, Captain S. W. DSC *The War at Sea 1939-1945* HMSO 1960.

de Sarrepont, H. (pseud) *Les Torpilles* Librairie Militaire de L. Baudoin, Paris 1883.

Schmalenbach, Paul 'The Genealogy of the Schnellboot' Warship International, Winter 1969, vol. 6.

Schofield, Vice-Admiral B. B. CB CBE *Operation Neptune* Ian Allan Ltd, Shepperton-on-Thames 1974.

Scott, Lieutenant Commander Peter MBE DSC & Bar *The Battle of the Narrow Seas* Country Life Ltd, London 1945.

Scott-Paine, H. 'Some of the Aspects and Problems of the Development of High-Speed Craft and its Machinery', Institution of Mechanical Engineers January 1939, *Proceedings* Vol. 141.

Sharphouse, R. P. 'The Use of Timber in Small Craft Construction', Royal Institution of Naval Architects, *Symposium on small craft* 1971.

Sharples, A. K. & McLeod, J. D. 'Pilot Launches – Design & Operation', Royal Institution of Naval Architects, *Symposium on small craft* 1971.

Silverstone, Paul H. *US Warships of World War II* Ian Allan Ltd, Shepperton-on-Thames 1965.

Sleeman, G. *Torpedoes and Torpedo Warfare* Griffin & Co, Portsmouth 1880 & 1889 eds.

Smith, Eng Captain Edgar C. OBE *A Short History of Naval and Marine Engineering* Cambridge University Press, London 1938.

Smith, Peter *Pedestal* William Kimber, London 1970.

Smith, Peter *Hard Lying* William Kimber, London 1971.

Spek, J. Dutch MTBs in the Japanese Navy, 'Warship International' Fall 1969.

Spyer, A. 'The Machinery of Small Boats for Ships of War', Institution of Naval Architects, March 1887, *Transactions* vol. 28.

Sueter, Commander Murray *The Evolution of the Submarine Boat, Mine and Torpedo* Griffin & Co, Portsmouth 1907.

Taylor, J. C. *German Warships of World War II* Ian Allan Ltd, Shepperton-on-Thames 1968.

Terry, C. Sanford (ed.) *Ostend and Zeebrugge Dispatches of Vice-Admiral Sir Roger Keyes KCB* Oxford University Press, London, 1919.

Thornycroft, Sir J. E. & Bremner, Lieutenant W. DSO 'Coastal Motor Boats: their design and service during the war', Institution of Naval Architects March 1923, *Transactions* vol. 65.

Thornycroft, Sir J. E. 'Torpedo Boats', an address to the Institution of Mechanical Engineers March 1937, *Proceedings* vol. 136.

Thornycroft, J. I. & Barnaby, S.W.'Torpedo-Boat Destroyers', Institution of Civil Engineers 1895, *Proceedings* vol. 122.

Thornycroft, J. I. & Co Ltd *A short History of the Revival of the Small Torpedo Boat (CMBs) during the Great War,* London 1918.

Trager, Frank N. *Why Vietnam?* Pall Mall Press, London 1966.

Tregaskis, Richard *John F. Kennedy & 'PT. 109'* Random Publications (After the Battle), Plaistow, 1975.

United States Naval History Division *Riverine Warfare* United States Navy, Washington 1968.

Van de Kasteele, Lieutenant W. J. 'The MTB: its Capabilities & Possibilities', Royal United Services Institute 1938, *Journal* vol. 83.

di Villarey, Captain Charles D. 'The work of the Italian Navy in the Adriatic during the War', Royal United Services Institute March 1919, *Journal* vol. 64.

Von Schertel, Hanns; Faber, Egon Dipling & Schatte, Eugen Dipling 'Military Hydrofoils' *Jane's Surface Skimmers 1972-1973,* Macdonald & Jane's, London 1972.

Watts, Anthony J. *Japanese Warships of World War II* Ian Allan Ltd, Shepperton-on-Thames 1966.

Watts, Anthony J. & Gordon, Brian *The Imperial Japanese Navy* Macdonald, London 1971.

White, W. L. *They Were Expendable* Harcourt, Brace & Co, New York 1943.

Wigzell, Halsey & Co *Greek-Fire Torpedo Boats* Smart & Allen, London 1877.

Wingate, John DSC HMS *Campbeltown* Profile Publications, Windsor 1971.

Wolter, Gustav-Adolf *See und Seefahrt* Koehlers Verlagsgesellschaft, Herford 1968.

Woodward, David *The Secret Raiders* William Kimber, London 1955.

JOURNALS and ANNUALS:
Motor Boat & Yachting; Mariner's Mirror; Engineering; Jane's Fighting Ships; Jane's Surface Skimmers; Flottes de Combat; Motor Boating.

INDEX